Functional Oxide Thin Films and Nanostructures: Growth, Properties, and Applications

Functional Oxide Thin Films and Nanostructures: Growth, Properties, and Applications

Editors

Paolo Mele
Satoru Kaneko
Tamio Endo

MDPI • Basel • Beijing • Wuhan • Barcelona • Belgrade • Manchester • Tokyo • Cluj • Tianjin

Editors

Paolo Mele
Shibaura Institute of
Technology (Tokyo)
Japan

Satoru Kaneko
Kanagawa Institute of
Industrial Science and
Technology (KISTEC)
Japan

Tamio Endo
Japan Advanced Chemicals
Japan

Editorial Office
MDPI
St. Alban-Anlage 66
4052 Basel, Switzerland

This is a reprint of articles from the Special Issue published online in the open access journal *Coatings* (ISSN 2079-6412) (available at: https://www.mdpi.com/journal/coatings/special_issues/oxide_thin_film_nanostruct).

For citation purposes, cite each article independently as indicated on the article page online and as indicated below:

LastName, A.A.; LastName, B.B.; LastName, C.C. Article Title. *Journal Name* **Year**, *Volume Number*, Page Range.

ISBN 978-3-0365-5657-4 (Hbk)
ISBN 978-3-0365-5658-1 (PDF)

© 2022 by the authors. Articles in this book are Open Access and distributed under the Creative Commons Attribution (CC BY) license, which allows users to download, copy and build upon published articles, as long as the author and publisher are properly credited, which ensures maximum dissemination and a wider impact of our publications.

The book as a whole is distributed by MDPI under the terms and conditions of the Creative Commons license CC BY-NC-ND.

Contents

About the Editors vii

Paolo Mele, Satoru Kaneko and Tamio Endo
Special Issue "Functional Oxide Thin Films and Nanostructures: Growth, Properties, and Applications"
Reprinted from: *Coatings* 2022, *12*, 778, doi:10.3390/coatings12060778 1

Musa Mutlu Can, Chasan Bairam, Seda Aksoy, Dürdane Serap Kuruca, Satoru Kaneko, Zerrin Aktaş and Mustafa Oral Öncül
Effect of Ti Atoms on Néel Relaxation Mechanism at Magnetic Heating Performance of Iron Oxide Nanoparticles
Reprinted from: *Coatings* 2022, *12*, 481, doi:10.3390/coatings12040481 3

Musa Mutlu Can, Yeşim Akbaba and Satoru Kaneko
Synthesis of Iron Gallate ($FeGa_2O_4$) Nanoparticles by Mechanochemical Method
Reprinted from: *Coatings* 2022, *12*, 423, doi:10.3390/coatings12040423 13

Ping-Yuan Lee, Endrika Widyastuti, Tzu-Che Lin, Chen-Tien Chiu, Fu-Yang Xu, Yaw-Teng Tseng and Ying-Chieh Lee
The Phase Evolution and Photocatalytic Properties of a $Ti-TiO_2$ Bilayer Thin Film Prepared Using Thermal Oxidation
Reprinted from: *Coatings* 2021, *11*, 808, doi:10.3390/coatings11070808 23

Saima Afroz Siddiqui, Deshun Hong, John E. Pearson and Axel Hoffmann
Antiferromagnetic Oxide Thin Films for Spintronic Applications
Reprinted from: *Coatings* 2021, *11*, 786, doi:10.3390/coatings11070786 41

Kamatam Hari Prasad, Karuppiah Deva Arun Kumar, Paolo Mele, Arulanandam Jegatha Christy, Kugalur Venkidusamy Gunavathy, Sultan Alomairy and Mohammed Sultan Al-Buriahi
Structural, Magnetic and Gas Sensing Activity of Pure and Cr Doped In_2O_3 Thin Films Grown by Pulsed Laser Deposition
Reprinted from: *Coatings* 2021, *11*, 588, doi:10.3390/coatings11050588 49

Veeraswamy Yaragani, Hari Prasad Kamatam, Karuppiah Deva Arun Kumar, Paolo Mele, Arulanandam Jegatha Christy, Kugalur Venkidusamy Gunavathy, Sultan Alomairy and Mohammed Sultan Al-Buriahi
Erratum: Yaragani et al. Structural, Magnetic and Gas Sensing Activity of Pure and Cr Doped In_2O_3 Thin Films Grown by Pulsed Laser Deposition. *Coatings* 2021, *11*, 588
Reprinted from: *Coatings* 2021, *11*, 1121, doi:10.3390/coatings11091121 63

Chien-Yie Tsay and Wan-Yu Chiu
Enhanced Electrical Properties and Stability of P-Type Conduction in ZnO Transparent Semiconductor Thin Films by Co-Doping Ga and N
Reprinted from: *Coatings* 202, *10*, 1069, doi:10.3390/coatings10111069 65

Hsin-Ming Cheng
Influence of the Growth Ambience on the Localized Phase Separation and Electrical Conductivity in $SrRuO_3$ Oxide Films
Reprinted from: *Coatings* 2019, *9*, 589, doi:10.3390/coatings9090589 79

Sara Massardo, Alessandro Cingolani and Cristina Artini
High Pressure X-ray Diffraction as a Tool for Designing Doped Ceria Thin Films Electrolytes
Reprinted from: *Coatings* **2021**, *11*, 724, doi:10.3390/coatings11060724 . . . . **89**

Mannarsamy Anitha, Karuppiah Deva Arun Kumar, Paolo Mele, Nagarajan Anitha, Karunamoorthy Saravanakumar, Mahmoud Ahmed Sayed, Atif Mossad Ali and Lourdusamy Almalraj
Synthesis and Properties of p-Si/n-$Cd_{1-x}Ag_xO$ Heterostructure for Transparent Photodiode Devices
Reprinted from: *Coatings* **2021**, *11*, 425, doi:10.3390/coatings11040425 . . . . **109**

Yan Tang, Yuxiang Zhang, Guanshun Xie, Youxiong Zheng, Jianwei Yu, Li Gao and Bingxin Liu
Construction of Rutile-TiO_2 Nanoarray Homojuction for Non-Contact Sensing of TATP under Natural Light
Reprinted from: *Coatings* **2020**, *10*, 409, doi:10.3390/coatings10040409 . . . . **123**

About the Editors

Paolo Mele

Paolo Mele (Professor) is currently a Professor at SIT Research laboratories, the Shibaura Institute of Technology, Tokyo, Japan. He obtained a master's degree in chemistry and a Ph.D. in chemical sciences at Genova University (Italy). In 2003, he moved to ISTEC-SRL in Tokyo to study melt-textured ceramic superconductors. Then, he worked as a postdoc at Kyoto University (JSPS fellowship) from 2004 to 2007, at the Kyushu Institute of Technology (JST fellowship) from 2007 to 2011, at Hiroshima University (as a lecturer) from 2011 to 2014, and at the Muroran Institute of Technology (as an associate professor) from 2015 to 2018 before his current position. His research interests include materials for energy and sustainable development (superconductors and thermoelectrics); the fabrication and characterization of thin films of oxides, ceramics, and metals; the study of the effect of nanostructuration on physical properties; thermal transport; and vortex matter. He is the author of over 110 papers in international scientific journals and four book chapters, has two patents, and has contributed to hundreds of communications at international conferences. He co-edited nine books on superconductors, oxides, thin films, thermoelectrics, and other materials for energy production.

Satoru Kaneko

Satoru Kaneko (Doctor) received a B.S. from the Tokyo Metropolitan Univ., an M.S. from the Univ. of Arizona, and a Ph.D. from the Tokyo Institute of Technology. His study focuses on the synthesis of functional materials of oxides, superconductors, and graphite-related materials, and he is also interested in the fabrication of nanostructures, for example, the self-organization of periodic nanostructures by laser scanning. He has published more than 100 papers in reputable journals. He spends his weekends road biking and jogging, and enjoys making bacon and beer in his backyard.

Tamio Endo

Tamio Endo (Professor emeritus) holds a Ph.D. (Kyoto University, Japan) and an MSD (Gifu University, Japan). He is an Emeritus Professor at Mie University (Japan), a Special Researcher at Gifu University (Japan), and an Honorary Professor of Southwest Jiaotong University (China) and was a Visiting Researcher at the University of California, San Diego, in 1995 (USA). He currently works at Japan Advanced Chemicals in Kanagawa (Japan). His research interests include oxide thin films, heterostructures, plasma effects, and the bonding of polymer films. He has been part of many international academic projects such as the Japan–India Cooperative Science Program. He has been an organizer and plenary speaker at many international conferences, has given many guest talks at foreign universities , and is a representative of the Team Harmonized Materials (formerly Team Harmonized Oxides).

Editorial

Special Issue "Functional Oxide Thin Films and Nanostructures: Growth, Properties, and Applications"

Paolo Mele [1,2,*], Satoru Kaneko [3] and Tamio Endo [4]

1. College of Engineering, Innovative Global Program, Shibaura Institute of Technology, 307 Fukasaku, Minuma-ku, Saitama 337-8570, Japan
2. International Research Center for Green Electronics, Shibaura Institute of Technology, 3-7-5 Toyosu, Koto-ku, Tokyo 135-8548, Japan
3. Kanagawa Institute of Industrial Science and Technology (KISTEC), 705-1 Shimo-Imaizumi, Ebina 243-0435, Japan; satoru@kistec.jp
4. Japan Advanced Chemicals, 3007-4 Kamiechi, Atsugi 243-0801, Japan; endotamio@yahoo.co.jp
* Correspondence: pmele@shibaura-it.ac.jp

It has been almost three years since we enthusiastically accepted the offer to be guest editors for this Special Issue of *Coatings*, entitled "Functional Oxide Thin Films and Nanostructures: Growth, Properties, and Applications".

Recent materials nanotechnologies have introduced the possibility of fabricating oxide thin films on a nanometric level, and this possibility also applies to nanocomposites. In parallel, recent measurement technologies can supply characterizations of their unique properties arising from limited regions of surfaces and interfaces. This Special Issue provides an opportunity to share surface-related science and engineering topics on oxide thin films and nanocomposites in an interactive and interdisciplinary manner. The goal is to elucidate commonalities and differences between multilayer interfaces and nanocomposite grain boundaries.

This Special Issue of *Coatings* was intended as an effort to bridge the gap between materials science and the applications of oxide thin films and nanostructures.

Originally, the topics of interest included but were not limited to: novel technologies to fabricate oxide nanomaterials; flexible and mechanically rigid oxide materials; wide categories of functional oxides (semiconducting, superconducting, magnetic, ferroelectric, multiferroic, optical); the understanding of structures and properties of oxide materials effectively exhibiting the above functions; the similarities and differences between "normal thin films" and "ultrathin films and multilayers", influenced by surfaces and interfaces; and the similarities and differences between "normal composites" and "nanocomposites", influenced by larger and smaller grains.

The ten published papers reflect the original spirit of the Special Issue, ranging from nanoparticles [1,2] to thin films [3–8] to heterostructures [9] and homojunctions [10] and covering various aspects of oxide-materials preparation, characterization, and applications.

We declare this Special Issue closed and thank all the colleagues and all the editorial staff of *Coatings* for their great contributions and unflagging support.

Funding: This research received no external funding.

Institutional Review Board Statement: Not applicable.

Informed Consent Statement: Not applicable.

Data Availability Statement: Not applicable.

Conflicts of Interest: The authors declare no conflict of interest.

References

1. Can, M.M.; Bairam, C.; Aksoy, S.; Kuruca, D.S.; Kaneko, S.; Akta¸s, Z.; Öncül, M.O. Effect of Ti Atoms on Néel Relaxation Mechanism at Magnetic Heating Performance of Iron Oxide Nanoparticles. *Coatings* **2022**, *12*, 481. [CrossRef]
2. Can, M.M.; Akbaba, Y.; Kaneko, S. Synthesis of Iron Gallate (FeGa$_2$O$_4$) Nanoparticles by Mechanochemical Method. *Coatings* **2022**, *12*, 423. [CrossRef]
3. Lee, P.-Y.; Widyastuti, E.; Lin, T.-C.; Chiu, C.-T.; Xu, F.-Y.; Tseng, Y.-T.; Lee, Y.-C. The Phase Evolution and Photocatalytic Properties of a Ti-TiO$_2$ Bilayer Thin Film Prepared Using Thermal Oxidation. *Coatings* **2021**, *11*, 808. [CrossRef]
4. Siddiqui, S.A.; Hong, D.; Pearson, J.E.; Hoffmann, A. Antiferromagnetic Oxide Thin Films for Spintronic Applications. *Coatings* **2021**, *11*, 786. [CrossRef]
5. Yaragani, V.; Kamatam, H.P.; Deva Arun Kumar, K.; Mele, P.; Christy, A.J.; Gunavathy, K.V.; Alomairy, S.; Al-Buriahi, M.S. Structural, Magnetic and Gas Sensing Activity of Pure and Cr Doped In$_2$O$_3$ Thin Films Grown by Pulsed Laser Deposition. *Coatings* **2021**, *11*, 588, Erratum: *Coatings* **2021**, *11*, 1121. [CrossRef]
6. Tsay, C.-Y.; Chiu, W.-Y. Enhanced Electrical Properties and Stability of P-Type Conduction in ZnO Transparent Semiconductor Thin Films by Co-Doping Ga and N. *Coatings* **2020**, *10*, 1069. [CrossRef]
7. Cheng, H.-M. Influence of the Growth Ambience on the Localized Phase Separation and Electrical Conductivity in SrRuO$_3$ Oxide Films. *Coatings* **2019**, *9*, 589. [CrossRef]
8. Massardo, S.; Cingolani, A.; Artini, C. High Pressure X-ray Diffraction as a Tool for Designing Doped Ceria Thin Films Electrolytes. *Coatings* **2021**, *11*, 724. [CrossRef]
9. Anitha, M.; Deva Arun Kumar, K.; Mele, P.; Anitha, N.; Saravanakumar, K.; Sayed, M.A.; Ali, A.M.; Amalraj, L. Synthesis and Properties of p-Si/n-Cd$_{1-x}$Ag$_x$O Heterostructure for Transparent Photodiode Devices. *Coatings* **2021**, *11*, 425. [CrossRef]
10. Tang, Y.; Zhang, Y.; Xie, G.; Zheng, Y.; Yu, J.; Gao, L.; Liu, B. Construction of Rutile-TiO2 Nanoarray Homojuction for Non-Contact Sensing of TATP under Natural Light. *Coatings* **2020**, *10*, 409. [CrossRef]

Article

Effect of Ti Atoms on Néel Relaxation Mechanism at Magnetic Heating Performance of Iron Oxide Nanoparticles

Musa Mutlu Can [1,*], Chasan Bairam [1], Seda Aksoy [2], Dürdane Serap Kuruca [3], Satoru Kaneko [4,5,6], Zerrin Aktaş [7] and Mustafa Oral Öncül [8]

1 Renewable Energy and Oxide Hybrid Systems Laboratory, Department of Physics, Faculty of Science, Istanbul University, Istanbul 34314, Turkey; chasanbairam@gmail.com
2 Department of Physics Engineering, Istanbul Technical University, Istanbul 34469, Turkey; sedaksoy@gmail.com
3 Department of Physiology, Faculty of Medicine, Istanbul University, Istanbul 34390, Turkey; sererdem@yahoo.com
4 National Cheng Kung University, Tainan 701, Taiwan; kaneko.s.al@m.titech.ac.jp
5 Electronics Engineering Department, Kanagawa Institute of Industrial Science and Technology (KISTEC), Ebina 243-0435, Kanagawa, Japan
6 Tokyo Institute of Technology, Nagatsuta, Yokohama 226-8502, Kanagawa, Japan
7 Faculty of Medicine, Department of Clinical Microbiology, Istanbul University, Istanbul 34390, Turkey; zaktas@istanbul.edu.tr
8 Faculty of Medicine, Department of Infectious Diseases and Clinical Microbiology, Istanbul University, Istanbul 34390, Turkey; oraloncul@yahoo.com
* Correspondence: musa.can@istanbul.edu.tr; Tel.: +090-(533)-929-0718

Abstract: The study was based on understanding the relationship between titanium (Ti) doping amount and magnetic heating performance of magnetite (Fe_3O_4). Superparamagnetic nanosized Ti-doped magnetite ((Fe_{1-x},Ti_x)$_3O_4$; x = 0.02, 0.03 and 0.05) particles were synthesized by sol-gel technique. In addition to (Fe_{1-x},Ti_x)$_3O_4$ nanoparticles, SiO_2 coated (Fe_{1-x},Ti_x)$_3O_4$ nanoparticles were produced as core-shell structures to understand the effects of silica coating on the magnetic properties of nanoparticles. Moreover, the magnetic properties were associated with the Néel relaxation mechanism due to the magnetic heating ability of single-domain state nanoparticles. In terms of results, it was observed that the induced RF magnetic field for SiO_2 coated ($Fe_{0.97}$,$Ti_{0.03}$)$_3O_4$ nanoparticles caused an increase in temperature difference (ΔT), which reached up to 22 °C in 10 min. The ΔT values of SiO_2 coated ($Fe_{0.97}$,$Ti_{0.03}$)$_3O_4$ nanoparticles were very close to the values of uncoated Fe_3O_4 nanoparticles.

Keywords: oxide semiconductor; point defects; Néel relaxation; magnetic hyperthermia; superparamagnetic nanoparticles

1. Introduction

The crystal structure of magnetite (Fe_3O_4) has a spinel cubic structure with Fd-3m space group [1,2]. Fe_3O_4 crystal structures obtain tetrahedral and octahedral sublattices, occupied by Fe^{2+} and Fe^{3+} cations coordinated with 32 oxygen atoms [1,2]. Ferrimagnetic properties are governed by the coupling of cation spins in octahedral and tetrahedral sites [1,2]. Properties, such as low toxicity, suitable magnetic properties, and easy fabrication, make the ferrite particles suitable for hyperthermia usages [3]. Each magnetic domain in a magnetic nanoparticle is oriented along the direction of the externally applied magnetic field. Also, especially for single-domain magnetic nanoparticles, all particles can be oriented in the direction of the magnetic field (Brownian motion), and the magnetic domains inside a particle can rotate without rotating the magnetic particle (Neel motion) [4]. In both cases, due to the desire to be in line with the external magnetic field, alternating magnetic fields initiate heating in and around single-domain magnetic particles.

During the last few decades, magnetic hyperthermia has become a widely researched topic (especially for cancer treatments) [5–11]. In recent technological applications, ferrite nanoparticles have emerged as the most popular candidate for magnetic hyperthermia applications. The heating performance of magnetic nanoparticles is investigated mainly depending on size and magnetic domain number in the particle [12]. The heating performance of multi-domain magnetic nanoparticles is related to energy losses from magnetic hysteresis due to the domain wall motions and eddy currents induced in magnetic grains [12,13]. Unlike multidomain nanosized particles, the magneto-heating performance of single magnetic domain nanosized particles is mainly correlated with magnetic anisotropy, inter/intra particles interactions and homogeny distributed magnetic nanoparticles [14]. Magnetic relaxation is generally the dominant mechanism in magnetic anisotropy and thus in magneto heating performance of nanoparticles. The magneto-heating performance of single magnetic domain nanosized particles is associated with two mechanisms, Néel and Brownian relaxations [15–17]. Brownian motion is created by mechanical fluctuation that performing by an entire nanoparticle rotating its own axis. On the other hand, the Néel relaxation mechanism is independent from the rotational motion of particles. The internal flip of spins with respect to the crystalline lattice creates magnetic heating associated with Néel relaxation [18–20]. Both relaxation mechanisms cause an increase in the temperature of magnetic nanoparticles. Even though magneto-heating can be observed with temperature increase, the magneto-heating performance of magnetic nanoparticles can be measured by specific absorption rate (SAR) values. SAR values can be defined by absorbed/converted magnetic energy into thermal energy and in technological applications (especially in cancer treatments) mainly modified properties of magnetic nanoparticles [3]. For magnetic hyperthermia applications, Néel relaxation has certain advantages due to its high SAR values [3]. On the other hand, Brownian relaxation is largely dependent on the viscosity of the medium surrounded by particles, and in cancer treatment SAR value is not enough for magneto-heating processes [3]. Liquid environmental conditions, such as different viscosities of the medium, agglomeration of nanoparticles within different cells, or fixation of nanoparticles in cell membranes (or extracellular tissue), weaken or inhibit the mechanical rotation of magnetic nanoparticles [21]. During the cancer treatments, due to the lack of movement or fluctuation of magnetic nanoparticles inside cancer cells Néel relaxation became the dominant factor for magnetic heating in compare to Brownian movement [3,20,21].

The study is based on the evaluation of the magneto heating performance of individual, non-interacting, and monodisperse particles in highly viscous environments confirming the Néel relaxation mechanism. The magneto heating mechanism is investigated for superparamagnetic Fe_3O_4 nanoparticles doped with Ti atoms. The magneto heating performance of magnetic nanoparticles is not sufficient for cancer treatments, and therefore the surface of magnetic nanoparticles must be functionalized by ligands [22,23]. The general way to improve the surface modification and functionalization ability of magnetic nanoparticles is to coat each nanoparticle with a SiO_2 shell [23]. Optical transparency, highly biocompatibility, biodegradability, and manufacturability with porous surfaces make SiO_2 very useful as a biomaterial [24]. The study also includes understanding the effect of SiO_2 on the heating performance of $(Fe_{1-x},Ti_x)_3O_4$ nanoparticles. Although coating with SiO_2 has no effect on crystal structures, the SiO_2 thickness has caused a remarkable decrease in internal magnetization value of Fe_3O_4 [25]. One of the consequences of the reduction of internal magnetization can be observed as a decrease in the magneto heating performance of magnetic nanoparticles [26,27]. On the other hand, SiO_2 coating reduces interparticle interactions and inhibits aggregation, thereby increasing the heating capabilities of magnetic nanoparticles in AC fields [28–30]. Therefore, the magneto heating performance of magnetic nanoparticles is correlated with interacting or non-interacting particles [26–30].

The magnetic nanoparticles such as Fe_3O_4 seem as efficient nano-heaters in biomedical applications [31–33]. Surface modified $(Fe_{1-x},Ti_x)_3O_4$ nanoparticles are expected to be potential bio-materials which suitable for loading anticancer drugs and heating under both excitations, RF magnetic field, and UV radiation, which utilize the magnetic nanoparticles useful at clinical hyperthermia applications.

2. Experimental

Magnetite nanoparticles were prepared via the co-precipitation method. Ferrous chloride tetrahydrate (FeCl$_2$·4H$_2$O) and ferric chloride (FeCl$_3$) were used as iron precursors containing different valance states. On the other hand, Tetraisopropil Ortotinatate (TIPO) was used as a Titanium source. Natrium hydroxide (NaOH: 25% by weight) in H$_2$O and hydrochloric acid in water (HCl) were used as the precipitating agent. To prevent the agglomeration of magnetic nanoparticles, nanoparticles were coated with oleic acid at the end of the process. Ethanol and DI water were used to remove excessive coating agent. The schematic diagram of the procedure was demonstrated in Figure 1. The synthesis was performed on magnetic stirrer at 90 °C. Firstly, ferrous and ferric chloride iron salts were dissolved in water with HCl under Argon gas flow. Secondly, after half an hour, TIPO and NaOH were dropped into the solution. Oleic acid was added to the solution as the last step.

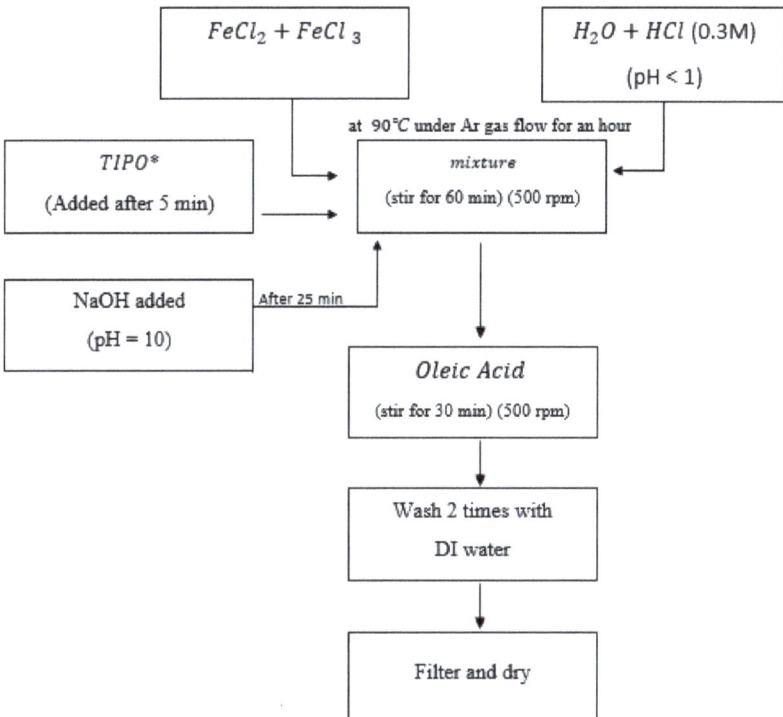

Figure 1. The schematic diagram of the synthesis procedure of pure and Ti-doped magnetite.

The coated nanoparticles were washed with DI water and ethanol to remove chloride ions and excessive coating materials. Furthermore, the same procedure was performed for the synthesis of the pure magnetite nanoparticles.

The nanoparticles were also coated with SiO$_2$ by base-catalyzed silica formation from tetraethylorthosilicate (TEOS) in a water-in-oil microemulsion technique, which was mentioned in previous study [34]. The resulting mixture was vigorously stirred for 24 h.

The crystal structures of samples were investigated by x-ray powder diffractometer (XRD), employing a MiniFlex model XRD, produced by Rigaku Corporation (Tokyo, Japan). The XRD patterns were taken under Cu K$_\alpha$ radiation (1.5406 Å) in 2θ range from 10°

to 90°. A JEM-2010F high-resolution transmission electron microscope (HR-TEM) (JEOL Ltd., Tokyo, Japan) was used to picture the structural morphologies of nanoparticles. Dc magnetization (σ (H)) measurements were performed at 300 K temperatures in a magnetic field range of $\pm$ 2 T.

Magneto-thermal characterization was taken by a homemade setup constructed using the equipment with a frequency of 150 kHz (power generator, thermometer, etc.). Experiments were performed in a custom-made setup with an alcohol thermometer, a covered glass tube, a water-cooled magnetic coil (diameter 50 mm and four turns for 160 Oe), and an AC power generator (Istanbul, Turkey) with a constant frequency of 150 kHz. The SAR measurements were conducted in non-adiabatic conditions as in many publications. The colloidal solution was put into a glass tube and the tube was placed in the coil. The thermometer was directly inserted into the solution. The temperature was measured using the thermometer as a function of time for a duration of 15 min. The filling level of the solution in the tube was adapted to the half-length of the coil to minimize the effects of magnetic field inhomogenity. Between the glass tube and the coils, we used styrofoam as an insulating material.

3. Result and Discussions

The structural analyses were performed employing XRD patterns for SiO_2 coated and uncoated particles as shown in Figure 2a,b respectively. The patterns were in an agreement with Fe_3O_4 diffraction pattern shown in ICDD card (PDF# 74-0748). No contamination or unexpected phase such as TiO_2 based structures, was detected on the XRD patterns. As seen in Figure 2b, even though having high background intensity, originating from the amorphous phase of SiO_2, the Fe_3O_4 patterns were distinctly distinguished at each XRD pattern.

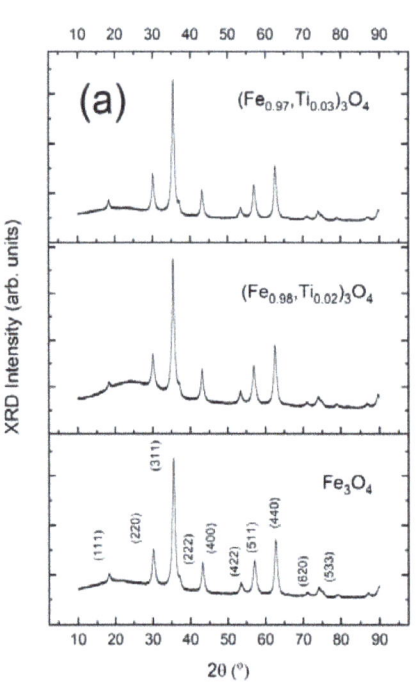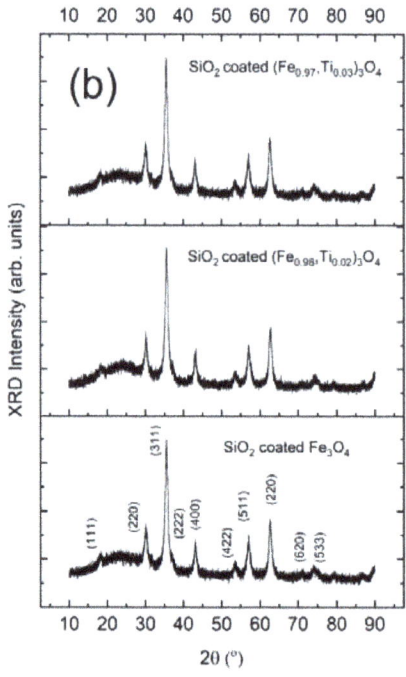

Figure 2. The XRD patterns of (**a**) Ti-doped magnetite and (**b**) SiO_2 coated Ti-doped magnetite.

Not observing Ti elements or compounds in the xrd patterns indicated the displacement of Ti atoms inside of Fe_3O_4 lattice. The ionic radius of Ti^{4+} is approximately 0.61Å, which is close to the ionic radius of Fe^{3+} (0.64 Å). Thus, Ti^{4+} ions are expected to settle instead of by Fe^{3+} ions in the octahedral lattice sites. Due to charge neutrality, a Ti^{4+} ion replacement in an octahedral site gives rise to change the valence state of Fe^{3+} ion to Fe^{2+} ion as shown in the chemical equation of (1) [35]. Due to its charge neutrality, the substitution of Ti^{4+}, a tetravalent positive ion, causes an Fe^{3+} ion to change its valency into Fe^{2+}. Chemical Equations (1) and (2) assign the charge neutrality occurring in $(Fe_{1-x},Ti_x)_3O_4$ lattice.

$$Fe^{2+} \leftrightarrow Fe^{3+} + e^- \tag{1}$$

$$2Fe^{3+} \rightarrow Ti^{4+} + Fe^{2+} \tag{2}$$

Increase in Ti^{4+} substitution amount ($0.01 \leq x \leq 0.025$) in $(Fe_{1-x},Ti_x)_3O_4$ lattice cause to formation of $Fe^{2+}{}_A(Fe^{2+}, Ti^{4+})_BO_4$ (A, tetrahedral side; B, octahedral side), which inhibits the hopping mechanism between iron ionic states. And thus, the new configuration cause to increase in magnetic anisotropy as mentioned in literature [35,36]. Excessive Ti^{4+} substitution ($x \geq 0.025$) easily bypasses Fe^{2+} ions, resulting in the formation of differential vacancies [35].

After understanding the crystal structure and possible defects in a lattice, the particle size distributions were investigated employing TEM micrographs. TEM Figures assigned homogeny distributed nanoparticles. In addition, due to covering with oleic acid, no agglomeration across the entire particle distribution was realized as seen in Figure 3.

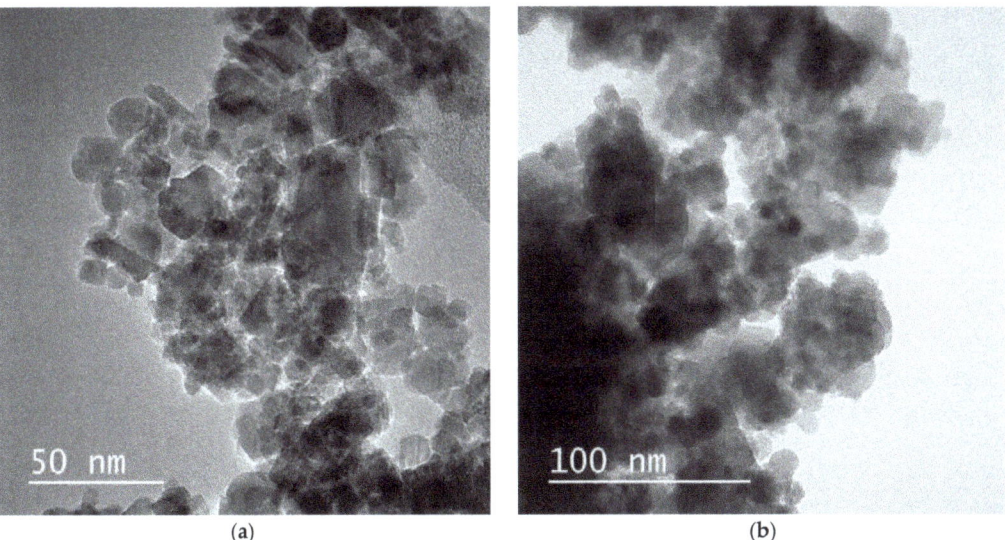

Figure 3. TEM micrographs of (**a**) uncoated and (**b**) coated $(Fe_{0.97},Ti_{0.03})_3O_4$ nanoparticles.

The size frequencies of particles were calculated by the subprogram of Image-J2. Particles with high differences in particle size were selected in the calculations. As shown on Figure 4, the particles size distributions which were confirmed by ImageJ, were found as 10.3 ± 0.6 nm and 18.7 ± 0.5 nm for uncoated and coated $(Fe_{0.97},Ti_{0.03})_3O_4$, respectively. The particles' size distribution indicated that the particles were in superparamagnetic regions.

The magnetization (σ (H)) measurements at the room temperature were illustrated in the Figure 5 for both samples uncoated and coated $(Fe_{1-x},Ti_x)_3O_4$. As seen from the Figure, the zero remanent magnetization assign the overcoming thermal energy to the magnetic anisotropy energy barrier at all samples. For both coated and uncoated $(Fe_{1-x},Ti_x)_3O_4$

particles, only difference in magnetization curves was the decrease in magnetization value by Ti amount in lattice. In order to understand the magnetic domain states of particles, room temperature magnetic hysteresis curves were obtained as shown in Figure 5.

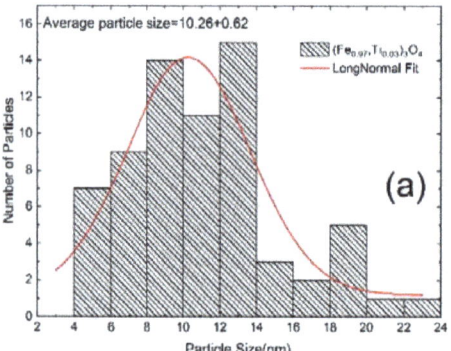
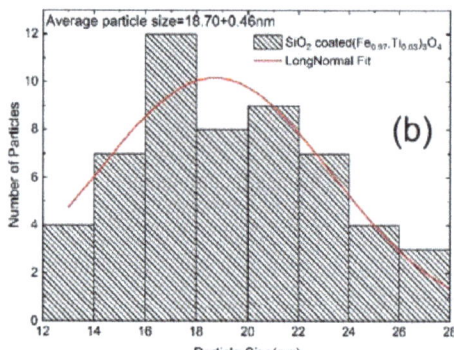

Figure 4. Calculated size distributions of (**a**) uncoated and (**b**) coated $(Fe_{0.97},Ti_{0.03})_3O_4$ nanoparticles via ImageJ subprogram.

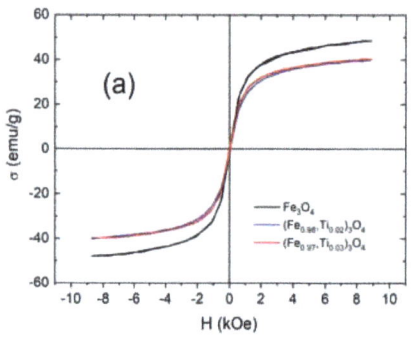
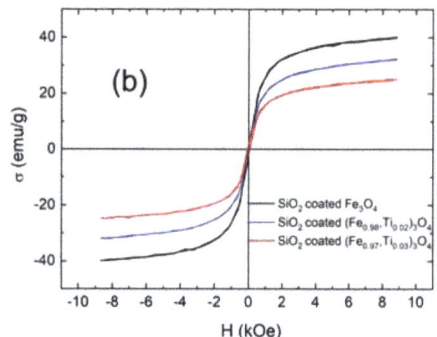

Figure 5. The magnetization measurements of (**a**) uncoated and (**b**) SiO_2 coated pure and Ti-doped magnetite.

Not observing coercivity and remanence values on hysteresis curves at room temperature proved that the particles were in superparamagnetic states [37,38]. Heating performance of an individual, non-interacting, and monodisperse particles have high SAR value due to dominating of Néel relaxation in a highly viscous environment. In the superparamagnetic state at room temperature, the magnetic interaction decreased to the lowest values with the effect of thermal energy. Nanoparticle size, anisotropy and interparticle interaction are the principal factors that influence heat generation [39,40]. Moreover, agglomerations between the magnetic nanoparticles occur due to strong magnetic dipole–dipole interactions between particles. As seen in Figure 5, coating with SiO_2 and doping with Ti atoms reduced the magnetization of the particles. A possible mechanism reducing the magnetization of particles is the number of vacancies in the crystal structure. The vacancies may form non-magnetic regions on the particle surface [41]. In addition, the magnetic anisotropy of the easy-axis is another parameter affecting the magnetization values that can be different either parallel or perpendicular to the easy-axis orientation of the domains [42]. These parameters manage the SAR values of particles.

Being predominant of thermal energy to magnetic energy at room temperature indicated that the nanoparticles were in superparamagnetic region. Since the particles were in superparamagnetic state size, the Néel magnetic relaxation was expected to dominate the magneto heating performance of the particles.

In Figure 6, the heating performance of particles was investigated under an ac magnetic field, approximately 13 kA/m field intensity and frequency of 150 kHz (the biological limits are 5×10^9 A/(m.s). The magneto heating measurements were taken immediately after arranging nanoparticles as magnetic fluids in 1 ml ethanol media. As seen on the Figure, the temperature difference (ΔT) reach up to 30 °C for pure Fe_3O_4 and for the same time interval Ti doping lowered the ΔT value down to 20 °C (($Fe_{0.97},Ti_{0.03})_3O_4$ nanoparticles). For ($Fe_{0.97},Ti_{0.03})_3O_4$ nanoparticles, the ΔT value was measured as 22 °C, which was good enough for use in vivo studies.

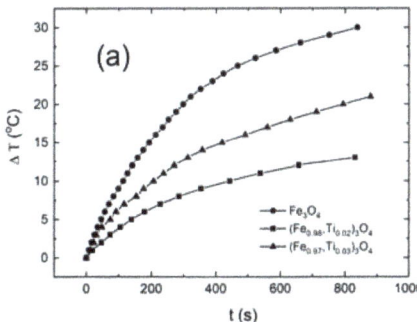 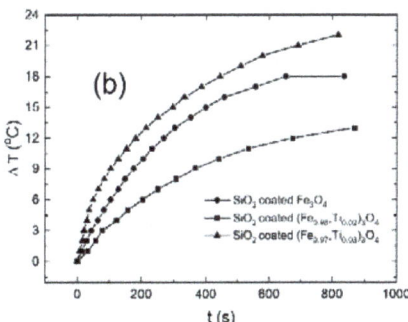

Figure 6. The magnetic heating performance of (**a**) uncoated and (**b**) SiO_2 coated pure and Ti doped Fe_3O_4.

Then, for each particle, the SAR value was calculated as shown in Table 1. The physical quantity of SAR value was determined by defined as the heat released from colloidal magnetic nanoparticles in unit time by Equation (3) [43].

$$SAR = \frac{Q}{\Delta t \, m_{mag}} \quad (3)$$

where $Q = mc\Delta T$, m_{mag} is the mass of magnetic nanoparticle, c is the specific heat of the colloid (only ethanol is taken into account, the contribution of magnetic nanoparticles, oleic acid, and SiO_2 to the specific heat are neglected). The calculations were performed for the heat capacity and the density of ethanol 2.57 kJ/(kgK), 0.789 g/mL, respectively.

Table 1. SAR values of $(Fe_{1-x},Ti_x)_3O_4$ (x = 0.00, 0.02 and 0.03) nanoparticles.

Nanoparticles	SAR (W/g)
Fe_3O_4	155
x = 0.02	70
x = 0.03	3
SiO_2 coated Fe_3O_4	104
x = 0.02 (SiO_2 coated)	34
x = 0.03 (SiO_2 coated)	116

The SAR values were illustrated in Table 1. As understood from Table 1, coating with SiO_2 lowered the SAR values of $(Fe_{1-x},Ti_x)_3O_4$ nanoparticles. However, for SiO_2 coated samples increase in Ti^{4+} ions amount in lattice caused an increase in SAR value, which getting closer to the value of pure Fe_3O_4 nanoparticles.

4. Conclusions

In the study, homogeny size distributed $(Fe_{1-x},Ti_x)_3O_4$ ferrite nanoparticles in oleic acid and at SiO_2 matrix were synthesized via a chemical route. The particles were obtained as superparamagnetic Fe_3O_4 nanoparticles to dominate the Néel relaxation over Brownian relaxation mechanism. Furthermore, lowering the particle size down to superparamagnetic region, coating with SiO_2 and Ti doping into the lattice was the tuned parameters to produce individual, non-interacting, and monodisperse particles. Then, the heating mechanism of SiO_2 coated Ti doped Fe_3O_4 nanoparticles were only correlated with Ti atoms amount in the lattice. Due to the expected coupling between Ti^{4+}-Fe^{2+} ions in the octahedral site, the heating performance by Ti doping was lower than pure Fe_3O_4. On the other hand, for SiO_2 coated $(Fe_{0.97},Ti_{0.03})_3O_4$ nanoparticles, the increase in the amount of Ti^{4+} ions in lattice cause an increase in SAR value ($\Delta T = 22$ °C in 10 min), while decreasing for uncoated nanoparticles. The heating performance of $(Fe_{0.97},Ti_{0.03})_3O_4$ nanoparticles coated with SiO_2 was almost close to the heating performance of pure magnetite.

Author Contributions: Conceptualization, M.M.C.; methodology, M.M.C.; formal analysis, M.M.C.; investigation, M.M.C., D.S.K., Z.A. and M.O.Ö.; resources, M.M.C.; data curation, M.M.C., C.B. and S.A.; writing—original draft preparation, M.M.C.; writing—review and editing, M.M.C., S.A., D.S.K., S.K., Z.A. and M.O.Ö.; visualization, M.M.C.; supervision, M.M.C.; project administration, M.M.C.; All authors have read and agreed to the published version of the manuscript.

Funding: This work was also supported by Scientific Research Projects Coordination Unit of Istanbul University with project number FBG-2018-28289.

Institutional Review Board Statement: Not applicable.

Informed Consent Statement: Not applicable.

Data Availability Statement: Data sharing is not applicable to this article.

Conflicts of Interest: The authors declare no conflict of interest.

References

1. Can, M.M.; Coşkun, M.; Fırat, T. Domain state-dependent magnetic formation of Fe_3O_4 nanoparticles analyzed via magnetic resonance. *J. Nanopart Res.* **2011**, *13*, 5497. [CrossRef]
2. Can, M.M.; Ozcan, S.; Fırat, T. Magnetic behaviour of iron nanoparticles passivated by oxidation. *Phys. Stat. Sol. C* **2006**, *3*, 1271–1278. [CrossRef]
3. Fortin, J.-P.; Gazeau, F.; Wilhelm, C. Intracellular heating of living cells through Néel relaxation of magnetic nanoparticles. *Eur. Biophys. J.* **2008**, *37*, 223–228. [CrossRef] [PubMed]
4. Jeyasubramanian, K.; Selvakumar, N.; Ilakkiya, J.; Santhoshkumar, P.; Satish, N.; Sahoo, S.K. Magnetic Flux Alignment Studies on Entrapped Ferrofluid Nanoparticles in Poly Vinyl Alcohol Matrix. *J. Mater. Sci. Technol.* **2013**, *29*, 903–908. [CrossRef]
5. Hilger, I.; Hergt, R.; Kaiser, W.A. Towards breast cancer treatment by magnetic heating. *J. Magn. Magn. Mater.* **2005**, *293*, 314–319. [CrossRef]
6. Nikam, D.S.; Jadhav, S.V.; Khot, V.M.; Phadatare, M.R.; Pawar, S.H. Study of AC magnetic heating characteristics of $Co_{0.5}Zn_{0.5}Fe_2O_4$ nanoparticles for magnetic hyperthermia therapy. *J. Magn. Magn. Mater.* **2014**, *349*, 208–213. [CrossRef]
7. Ortega, D.; Pankhurst, Q.A. *Magnetic Hyperthermia, in Nanoscience: Volume 1: Nanostructures through Chemistry*; O'Brien, P., Ed.; Royal Society of Chemistry: Cambridge, UK, 2013; pp. 60–88.
8. Périgo, E.A.; Hemery, G.; Sandre, O.; Ortega, D.; Garaio, E.; Plazaola, F.; Teran, F.J. Fundamentals and advances in magnetic hyperthermia. *Appl. Phys. Rev.* **2015**, *2*, 041302. [CrossRef]
9. Suto, M.; Hirota, Y.; Mamiya, H.; Fujita, A.; Kasuya, R.; Tohji, K.; Jeyadevan, B. Heat dissipation mechanism of magnetite nanoparticles in magnetic fluid hyperthermia. *J. Magn. Magn. Mater.* **2009**, *321*, 1493–1496. [CrossRef]
10. Obaidat, I.M.; Issa, B.; Haik, Y. Magnetic Properties of Magnetic Nanoparticles for Efficient Hyperthermia. *Nanomaterials* **2015**, *5*, 63–89. [CrossRef]
11. Dennis, C.L.; Ivkov, R. Physics of heat generation using magnetic nanoparticles for hyperthermia. *Int. J. Hyperth.* **2013**, *29*, 715–729. [CrossRef]
12. Verges, M.A.; Costo, R.; Roca, A.G.; Marco, J.F.; Goya, G.F.; Serna, C.J.; Morales, M.P. Uniform and water stable magnetite nanoparticles with diameters around the monodomain–multidomain limit. *J. Phys. D Appl. Phys.* **2008**, *41*, 134003. [CrossRef]
13. Skumiel, A.; Kaczmarek-Klinowska, M.; Timko, M.; Molcan, M.; Rajnak, M. Evaluation of Power Heat Losses in Multidomain Iron Particles under the Influence of AC Magnetic Field in RF Range. *Int. J. Thermophys.* **2013**, *34*, 655–666. [CrossRef]
14. Deatsch, A.E.; Evans, B.A. Heating efficiency in magnetic nanoparticle hyperthermia. *J. Magn. Magn. Mater.* **2014**, *354*, 163–172. [CrossRef]

15. Ilg, P.; Kröger, M. Dynamics of interacting magnetic nanoparticles: Effective behavior from competition between Brownian and Néel relaxation. *Phys. Chem. Chem. Phys.* **2020**, *22*, 22244–22259. [CrossRef] [PubMed]
16. Çelik, Ö.; Can, M.M.; Fırat, T. Size dependent heating ability of $CoFe_2O_4$ nanoparticles in AC magnetic field for magnetic nanofluid hyperthermia. *J. Nanopart. Res.* **2014**, *16*, 232. [CrossRef]
17. Ganguly, S.; Margel, S. Review: Remotely controlled magneto-regulation of therapeutics frommagnetoelastic gel matrices. *Biotechnol. Adv.* **2020**, *44*, 1076112. [CrossRef] [PubMed]
18. Fannin, P.C.; Charles, S.W. The study of a ferrofluid exhibiting both Brownian and Néel relaxation. *J. Phys. D Appl. Phys.* **1989**, *22*, 187. [CrossRef]
19. Fannin, P.C.; Charles, S.W. On the calculation of the Néel relaxation time in uniaxial single-domain ferromagnetic particles. *J. Phys. D Appl. Phys.* **1994**, *27*, 185. [CrossRef]
20. Hergt, R.; Dutz, S.; Zeisberger, M. Validity limits of the Néel relaxation model of magnetic nanoparticles for hyperthermia. *Nanotechnology* **2010**, *21*, 015706. [CrossRef]
21. Fabris, F.; Lima, E.; Biasi, E.; Troiani, H.E.; Mansilla, M.V.; Torres, T.E.; Pacheco, R.F.; Ibarra, M.R.; Goya, G.F.; Zysler, R.D.; et al. Controlling the dominant magnetic relaxation mechanisms for magnetic hyperthermia in bimagnetic core–shell nanoparticles. *Nanoscale* **2019**, *11*, 3164–3172. [CrossRef]
22. Cole, A.J.; Yang, V.C.; David, A.E. Cancer theranostics: The rise of targeted magnetic nanoparticles. *Trends Biotechnol.* **2011**, *29*, 323–332. [CrossRef]
23. Schladt, T.D.; Schneider, K.; Schild, H.; Tremel, W. Synthesis and bio-functionalization of magnetic nanoparticles for medical diagnosis and treatment. *Dalton Trans.* **2011**, *40*, 6315–6343. [CrossRef] [PubMed]
24. Marcelo, G.A.; Lodeiro, C.; Capelo, J.L.; Lorenzo, J.; Oliveira, E. Magnetic, fluorescent and hybrid nanoparticles: From synthesis to application in biosystems. *Mater. Sci. Eng. C* **2020**, *106*, 110104. [CrossRef] [PubMed]
25. Husain, H.; Hariyanto, B.; Sulthonul, M.; Klysubun, W.; Darminto, D.; Pratapa, S. Structure and magnetic properties of silica-coated magnetitenanoparticle composites. *Mater. Res. Express* **2019**, *6*, 86117.
26. Lemal, P.; Balog, S.; Geers, C.; Taladriz-Blanco, P.; Palumbo, A.; Hirt, A.M.; Rothen-Rutishauser, B.; Petri-Fink, A. Heating behavior of magnetic iron oxide nanoparticles at clinically relevant concentration. *J. Magn. Magn. Mater.* **2019**, *474*, 637–642. [CrossRef]
27. Larumbe, S.; Gomez-Polo, C.; Perez-Landazabal, J.I.; Pastor, J.M. Effect of a SiO_2 coating on the magnetic properties of Fe_3O_4 nanoparticles. *J. Phys. Condens. Matter.* **2012**, *24*, 266007. [CrossRef]
28. Arteaga-Cardona, F.; Rojas-Rojas, K.; Costo, R.; Mendez-Rojas, M.A.; Hernando, A.; Presa, P. Improving the magnetic heating by disaggregating nanoparticles. *J. Alloy. Compd.* **2016**, *663*, 636–644. [CrossRef]
29. Ota, S.; Takemura, Y. Characterization of Néel and Brownian Relaxations Isolated from Complex Dynamics Influenced by Dipole Interactions in Magnetic Nanoparticles. *J. Phys. Chem. C* **2019**, *123*, 28859–28866. [CrossRef]
30. Kusigerski, V.; Illes, E.; Blanusa, J.; Gyergyek, S.; Boskovic, M.; Perovic, M.; Spasojevic, V. Magnetic properties and heating efficacy of magnesium doped magnetite nanoparticles obtained by co-precipitation method. *J. Magn. Magn. Mater.* **2019**, *475*, 470–478. [CrossRef]
31. Lak, A.; Disch, S.; Bende, P. Embracing Defects and Disorder in Magnetic Nanoparticles. *Adv. Sci.* **2021**, *8*, 2002682. [CrossRef]
32. Lavorato, G.C.; Das, R.; Masa, J.A.; Phan, M.-H.; Srikanth, H. Hybrid magnetic nanoparticles as efficient nano heaters in biomedical applications. *Nanoscale Adv.* **2021**, *3*, 867–888. [CrossRef]
33. Sharifianjazi, F.; Irani, M.; Esmaeilkhanian, A.; Bazli, L.; Asl, M.S.; WonJang, H.; Kim, S.Y.; Ramakrishna, S.; Shokouhimehr, M.; Varma, R.S. Polymer incorporated magnetic nanoparticles: Applications for agnetoresponsive targeted drug delivery. *Mater. Sci. Eng. B* **2021**, *272*, 115358. [CrossRef]
34. Coskun, M.; Can, M.M.; Duyar-Coskun, Ö.; Korkmaz, M.; Fırat, T. Surface anisotropy change of $CoFe_2O_4$ nanoparticles depending on thickness of coated SiO_2 shell. *J. Nanopart. Res.* **2012**, *14*, 1197. [CrossRef]
35. Walz, F.; Torres, L.; Bendimya, K.; Francisco, C.; Kronmuller, H. Analysis of magnetic after-effect spectra in titanium-doped magnetite. *Phys. Status Solidi* **1997**, *164*, 805. [CrossRef]
36. Kakol, Z.; Sabol, J.; Stickler, J.; Kozfowski, A.; Honig, J.M. Influence of titanium doping on the magneto crystalline anisotropy of magnetite. *Phys. Rev. B* **1994**, *49*, 12767–12772. [CrossRef]
37. Petracic, O. Superparamagnetic nanoparticle ensembles. *Superlattices Microstruct.* **2010**, *47*, 569–578. [CrossRef]
38. Kim, D.K.; Zhang, Y.; Voit, W.; Rao, K.V.; Muhammed, M. Synthesis and characterization of surfactant-coated superparamagnetic monodispersed iron oxide nanoparticles. *J. Magn. Magn. Mater.* **2001**, *225*, 30–36. [CrossRef]
39. Lasheras, X.; Insausti, M.; Fuente, J.M.; Muro, I.G.; Castellanos-Rubio, I.; Marcano, L.; Fernández-Gubieda, M.L.; Serrano, A.; Martín-Rodríguez, R.; Garaio, E.; et al. Mn-Doping level dependence on the magnetic response of $Mn_xFe_{3x}O_4$ ferrite nanoparticles. *Dalton Trans.* **2019**, *48*, 11480–11491. [CrossRef]
40. Usov, N.A.; Serebryakova, O.N.; Tarasov, V.P. Interaction Effects in Assembly of Magnetic Nanoparticles. *Nanoscale Res. Lett.* **2017**, *12*, 489. [CrossRef]
41. Köhler, T.; Feoktystov, A.; Petracic, O.; Kentzinger, E.; Bhatnagar-Schöffmann, T.; Feygenson, M.; Nandakumaran, N.; Landers, J.; Wende, H.; Cervellino, A.; et al. Mechanism of magnetization reduction in iron oxide nanoparticles. *Nanoscale* **2021**, *13*, 6965–6976. [CrossRef]

42. Yamaminami, T.; Ota, S.; Trisnanto, S.B.; Ishikawa, M.; Yamada, T.; Yoshida, T.; Enpuku, K.; Takemura, Y. Power dissipation in magnetic nanoparticles evaluated using the AC susceptibility of their linear and nonlinear responses. *J. Magn. Magn. Mater.* **2021**, *517*, 167401. [CrossRef]
43. Rosensweig, R.E. Heating magnetic fluid with alternating magnetic field. *J. Magn. Magn. Mater.* **2002**, *252*, 370–374. [CrossRef]

Article

Synthesis of Iron Gallate (FeGa$_2$O$_4$) Nanoparticles by Mechanochemical Method

Musa Mutlu Can [1,*], Yeşim Akbaba [1] and Satoru Kaneko [2,3,4]

1 Renewable Energy and Oxide Hybrid Systems Laboratory, Department of Physics, Faculty of Science, Istanbul University, Vezneciler, 34314 Istanbul, Turkey; yesimakbaba34@gmail.com
2 Department of Material Science and Engineering, National Cheng Kung University, Tainan City 701, Taiwan; kaneko.s.al@m.titech.ac.jp
3 Kanagawa Institute of Industrial Science and Technology (KISTEC), Atsugi 243-0292, Japan
4 Tokyo Institute of Technology, Yokohama 226-8502, Japan
* Correspondence: musa.can@istanbul.edu.tr; Tel.: +90-(533)-929-0718

Abstract: The study was focused on optimizing the procedure of synthesizing iron gallate (FeGa$_2$O$_4$) nanoparticles by mechanochemical techniques. Due to a lack of information in the literature about the sequence of synthesis procedures of FeGa$_2$O$_4$ structures, the study is based on the establishment of a recipe for FeGa$_2$O$_4$ synthesis using mechanochemical techniques. Rotation speed, grinding media, and milling durations were the optimized parameters. At the end of each step, the structure of the resulting samples was investigated using the X-ray diffraction (XRD) patterns of samples. At the end of the processes, the XRD patterns of the samples milled under an air atmosphere were coherent with the XRD pattern of the FeGa$_2$O$_4$ structure. XRD patterns were analyzed employing Rietveld refinements to determine lattice parameters under the assumption of an inverse spinel crystal formation. Furthermore, a fluctuation at band gap values in the range of 2.39 to 2.55 eV was realized and associated with the excess Fe atoms in the lattice, which settled as defects in the crystal structures.

Keywords: spinel oxide semiconductors; nanoparticle catalysts; inverse spinel; mechanochemical synthesis; gallates

Citation: Can, M.M.; Akbaba, Y.; Kaneko, S. Synthesis of Iron Gallate (FeGa$_2$O$_4$) Nanoparticles by Mechanochemical Method. *Coatings* 2022, 12, 423. https://doi.org/10.3390/coatings12040423

Academic Editor: Aivaras Kareiva

Received: 22 February 2022
Accepted: 17 March 2022
Published: 22 March 2022

Publisher's Note: MDPI stays neutral with regard to jurisdictional claims in published maps and institutional affiliations.

Copyright: © 2022 by the authors. Licensee MDPI, Basel, Switzerland. This article is an open access article distributed under the terms and conditions of the Creative Commons Attribution (CC BY) license (https://creativecommons.org/licenses/by/4.0/).

1. Introduction

Spinel oxide semiconductors are widely investigated materials due to their attractive performance in a variety of technological applications, such as photocatalytic fuel cells, biomaterials, and sensors [1–5]. The chemical formula of spinel oxides can be defined as a cubic structure with a formula of AB$_2$O$_4$, where A and B represent a divalent metallic cation at the tetrahedral site and tetravalent metallic cations at the octahedral sites, respectively [6,7].

Among the vast variety of spinel oxide semiconductors, iron gallate (FeGa$_2$O$_4$) has recently been considered an outstanding material due to its magnetic properties; piezoelectric, magnetoelectric, and magneto-optical performance; and cathodoluminescent features [8–12]. FeGa$_2$O$_4$ has a cubic spinel crystal structure with a space group of Fd-3m and a lattice parameter of a = 8.385 Å [13].

FeGa$_2$O$_4$ was firstly investigated due to its piezoelectric and magnetoelectric properties. Nowadays, FeGa$_2$O$_4$ structures show remarkable properties in Li-ion capacitors as anode materials or in biotechnological applications [14–16]. The technological phenomena of structures composed of FeGa$_2$O$_4$ mainly originate from its inverse or normal spinel structures [17,18]. Studies have indicated that FeGa$_2$O$_4$ can be formed as either a full inverse structure or a partial inverse structure. However, studies on FeGa$_2$O$_4$ structures still need further investigation since the synthesis of FeGa$_2$O$_4$ structures is quite difficult. In the literature, the number of studies on a systematic method for the synthesis of FeGa$_2$O$_4$

structures are very few, and the processes are not easy to follow [10,11,13,14,18–21]. This is why there is still not enough information on synthesis processes.

$FeGa_2O_4$ structures also seem very attractive for use in thick/thin films. $FeGa_2O_4$ oxide thin films seem to be suitable materials for photocathodes due to their physical properties such as transparency to visible light, p-type electrical conductivity, high enough chemical resistivity to an acidic environment, and tunable band edge by sunlight to the redox potential of water. According to the needs of photocatalytic materials, $FeGa_2O_4$ structures appear suitable for photocatalytic fuel cell applications either as a thin film or a sponge-like thick film with nanowalls [22,23].

Mechanochemical syntheses are widely operated techniques used to produce a variety of functional materials [24]. Charge ratio, rotation speed, milling environment, and duration are parameters that may manage structural formation [25]. The synthesis parameters of the mechanochemical techniques were adjusted to synthesize $FeGa_2O_4$ structures, and the crystal structure of the synthesized particles was investigated by employing Rietveld refinements.

2. Experimental

Mechanochemical techniques were employed to synthesize iron gallate ($FeGa_2O_4$) nanopowders. During the synthesis processes, the parameters of rotation speed and milling environments were optimized. At each process, the same amount of Fe (purity 99.9% Sigma Aldrich), Ga_2O_3 (purity 99.9% Sigma Aldrich), and deionized water (H_2O) was used as the starting materials, and an excessive amount of H_2O was added to the container as a lubricant material.

Each milling process was performed employing a Planetary Mono Mill pulverisette 6 (FRITSCH GmbH, Idar-Oberstein, Germany) inside of a stainless steel vial with 10 mm diameter stainless steel balls. Three separate procedures were performed. The samples were named as *FeGaO-01*, *FeGaO-02*, and *FeGaO-03*. The *FeGaO-01* sample was obtained by the rotation speed of 300 rpm. After every 12 h of milling, the container was opened to the atmosphere and again milled for another 12 h. At the end of 84 h of the milling procedure, $FeGa_2O_4$ particles were produced. For the *FeGaO-02* sample, the rotation speed was increased up to 360 rpm and, again, the vial was opened after every 12 h of milling. At the end of 36 h of milling, $FeGa_2O_4$ particles were produced and named *FeGaO-02*. For the final sample, without opening the chamber to the atmosphere, a continuous milling procedure was performed for 36 h at a rotation speed of 360 rpm to understand the necessity of opening the vial to the atmosphere after each 12 h of milling. The produced sample was named *FeGaO-03*. The procedures are summarized in Table 1.

Table 1. The mechanochemical parameters during the synthesis of the particles.

Sample Name	The Rotation Speed (Rpm)	The Milling Environments	Times of Open to Ambient Atmosphere	Charge Ratio (Powder to Ball Weight Ratio)	Duration (Hour)
FeGaO-01	300	Air	8	1:30	84
FeGaO-02	360	Air	3	1:30	36
FeGaO-03	360	released gas *	0	1:30	36

* indicates released gas during the milling process.

To determine the crystal structures, XRD patterns were recorded by a MiniFlex model X-ray powder diffractometer (XRD) produced by Rigaku Corporation, with Cu K_α radiation (1.5406 Å) in 2 θ range from 10 to 90°. The XRD patterns were obtained by a scan rate of 0.008°/min steps, which are suitable for Reitveld refinements. The FOOLPROOF and MATCH subprograms were utilized to obtain the structural parameters.

The bandgap calculations of each powder were investigated by a Shimadzu UV-2600 model UV-VIS spectrophotometer (Shimadzu Scientific Instruments, Kyoto, Japan). The UV-Vis spectra were taken in the wavelength range of 195 nm to 1400 nm. The bandgap calculations for powder samples were performed by the data of diffuse reflectance spectra (DRS). Then, the DRS data were analyzed by using the Kubelka–Munk transformation, employing the UvProbe 2.70 subprogram.

3. Results and Discussion

The procedures were mainly based on synthesizing $FeGa_2O_4$ particles by employing mechanochemical techniques, without annealing steps. In the beginning, the starting materials were put into the vial and the particles were milled for 12 h periods at a 300 rpm rotation speed. At the end of milling periods, the vial was opened to the atmosphere to release the H_2 gasses. The XRD pattern of the product was taken, as shown in Figure 1a. At the end of each milling period until the end of the completed 84 h milling process, the powders contained more than one compound. According to the JCPDS pdf cards, we assumed that the possible compounds in our samples were $FeGa_2O_4$, α- FeOOH, γ- FeOOH, Ga_2O_3, and Fe_3O_4. The peak positions of contamination compounds were identified by arrows on figures. At the end of 84 h of milling at 300 rpm, it was noticed that the XRD pattern of final powders was coherent with the $FeGa_2O_4$ diffraction pattern with the pdf card number of JCPDS 01-074-2229. On the figures, $FeGa_2O_4$ peak positions were identified and indexed. No contamination compound or element was detected on the final pattern. The final product was named *FeGaO-01*, as seen in Figure 1a.

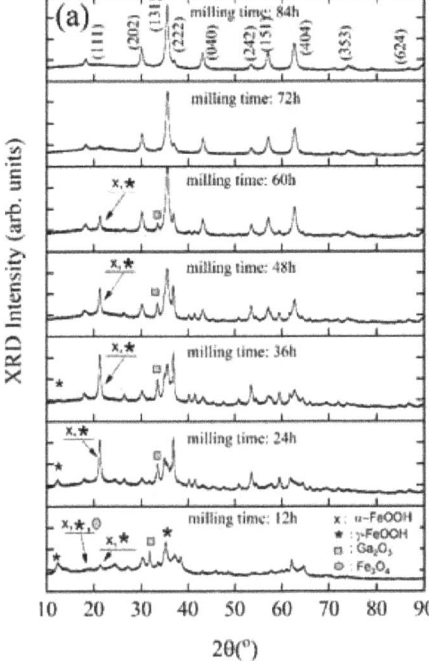
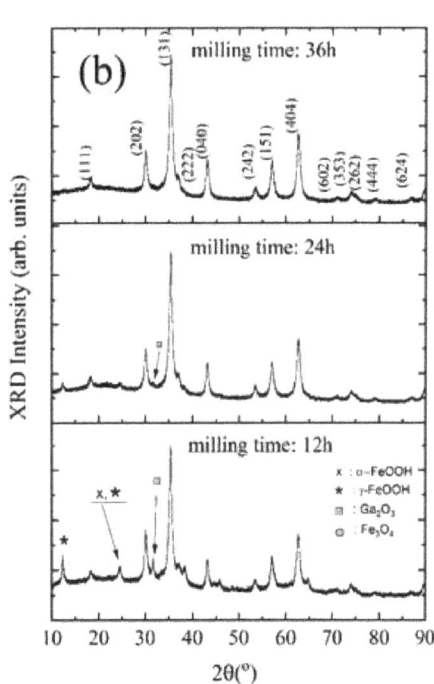

Figure 1. *Cont.*

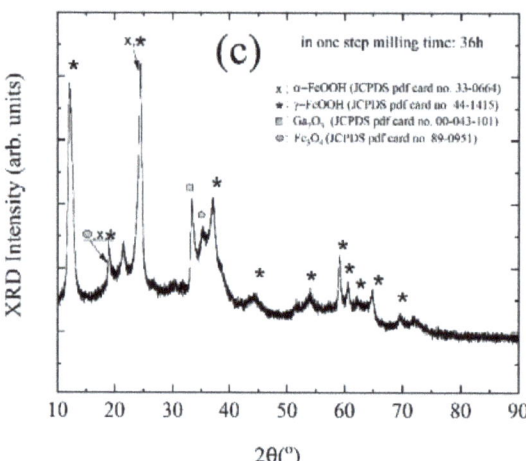

Figure 1. X-ray powder diffraction pattern of prepared samples by (**a**) a rotation speed of 300 rpm with 12 h milling periods, (**b**) a rotation speed of 360 rpm with 12 h milling periods, and (**c**) a rotation speed of 360 rpm without break in the milling. The diffraction patterns were indicated by the Joint Committee on Powder Diffraction Standards (JCPDS) powder diffraction file (pdf) card numbers, such as FeGa$_2$O$_4$, α—FeOOH, γ—FeOOH, Ga$_2$O$_3$, and Fe$_3$O$_4$, which were identified by 01-074-2229, 33-0664, 44-1415, 00-043-101, and 89-0951, respectively).

To decrease the milling time, the milling parameters were optimized and the *FeGaO-02* sample was obtained. For *FeGaO-02*, the rotation speed was increased to 360 rpm, and the milling time intervals were kept at 12 h. The XRD patterns of samples, produced at each step, are shown in Figure 1b. As seen from XRD patterns, repeating the milling process three times was enough to produce FeGa$_2$O$_4$ nanoparticles. Then, the process was performed at a 360 rpm rotation speed non-stop for 36 h, without a break in order to understand the need for taking a break. The sample produced without taking a break to release the H$_2$ gases out of vial was named *FeGaO-03*. The XRD pattern of *FeGaO-03* is shown in Figure 1c. As seen in Figure 1c, the milling without a break was not enough to synthesize FeGa$_2$O$_4$ nanopowders. The expected chemical reaction during the milling process can be shown by the chemical Equation (1):

$$Fe + Ga_2O_3 + H_2O \rightarrow FeGa_2O_4 + H_2 \tag{1}$$

As seen in the chemical reaction, H$_2$ gasses were expected to release into the vial. H$_2$ gasses can prevent the reaction from continuing, as observed for Fe$_3$O$_4$ particles that in situ produced H$_2$ and hindered the formation of a new crystal phase during milling [26]. During the milling process, opening the vial releases the H$_2$ gasses, which exit the vial and allow the formation of the chemical reaction under the ambient atmosphere to finalize as FeGa$_2$O$_4$.

During the milling processes, we expected a large strain to form in the powders. Broadening at the XRD peaks was expected as the indicator of the strain formed in the particles. That is why, for the powders, XRD peaks were associated with particle size distribution and the strain with the powders [27–31]. In general, the average particle size was calculated using Debye–Scherrer's formula, as shown in Equation (2), and the strain originating from crystal imperfection and distortion can be calculated by Equation (3). To obtain the average particle size distribution and understand the strain effects, a general formula was defined as the uniform deformation model (UDM). The UDM was defined by the equation shown in (4), which provides the effect of intrinsic strain associated with crystal size [28–31].

$$L = \frac{K\lambda}{\beta_{hkl}\cos\theta} \quad (2)$$

$$\varepsilon = \frac{\beta_{hkl}}{4\tan\theta} \quad (3)$$

$$\beta_{hkl}\cos\theta = \frac{K\lambda}{L} + 4\varepsilon\sin\theta \quad (4)$$

In the equation, the β_{hkl}, L, K, λ, ε, and θ symbols indicate the full width at the half-maximum of the diffracted peak, indicating the broadening of a peak, the average particle sizes, a constant defined as the shape factor (=0.9), the wavelength of the CuKα radiation (1.5406 Å), the strain, and the angle of diffraction, respectively.

From broadening XRD peaks, the particle sizes and strain associated with the samples of *FeGaO-01* and *FeGaO-02* were calculated by Equation (4), as shown in Figure 2a,b, respectively. The average particle size calculations were obtained from the diffraction peaks of (111), (202), (131), and (040). The crystallite sizes for the *FeGaO-01* and *FeGaO-02* samples were found as 14.4 nm and 12.0 nm, respectively. Furthermore, the microstrain of the particles was obtained by employing Equation (3), and the data are shown in Table 2. The microstrain values of the *FeGaO-01* and *FeGaO-02* samples, shown in Figure 2c,d, were calculated as $11.06 \pm 06 \times 10^{-2}$ and $1.1 \pm 0.2 \times 10^{-3}$, respectively. The microstrain is associated with a lattice dislocation density [32]. The longer time milling causes the formation of a 10 times higher microstrain with particles.

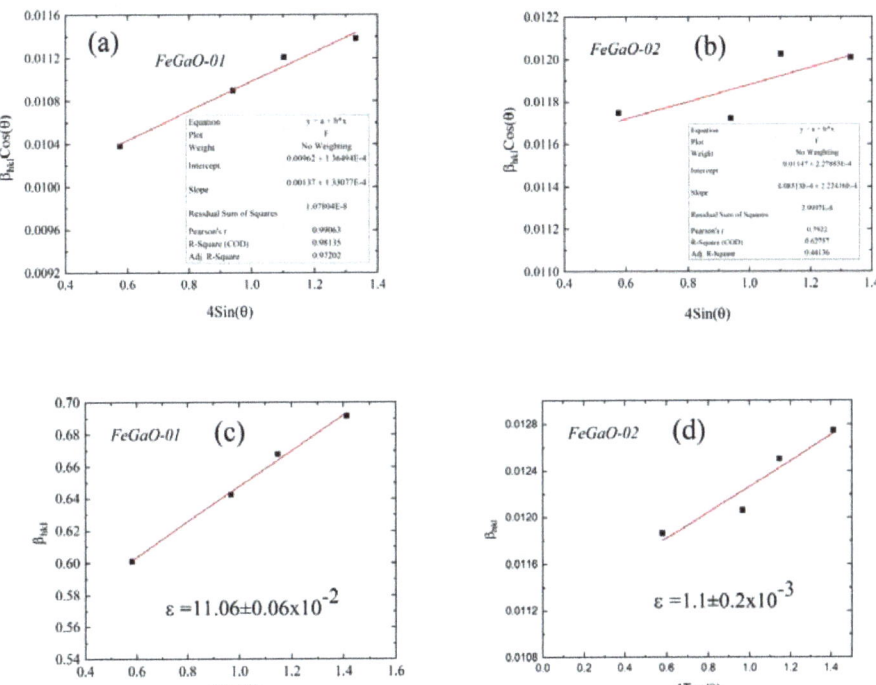

Figure 2. Fits to the uniform deformation model (UDM) for the samples to calculate the particle sizes of (**a**) *FeGaO-01* and (**b**) *FeGaO-02*, and to calculate the microstrain with particles of (**c**) *FeGaO-01* and (**d**) *FeGaO-02*.

Table 2. Rietveld refinement data and the band gap values of samples.

Sample Name	Particle Size (nm)	Microstrain of Particles	Rietveld Refinement				Occupancy Ratios					Band Gap (eV)
			Space Group	Bragg R Factors (R_B)	χ^2 Values	The Lattice Constant a (Å)	Tetrahedral Site			Octahedral Site		
							Fe1	Ga1	O3	Fe2	Ga2	
FeGaO-01	14.4	$11.06 \pm 0.06 \times 10^{-2}$	$F4d_132$ (No. 210)	0.842	0.368	8.3687 ± 0.0001	0.0693	0.1540	0.4417	0.1540	0.0693	2.39 ± 0.02
FeGaO-02	12.0	$1.1 \pm 0.2 \times 10^{-3}$		0.910	0.338	8.3790 ± 0.0003	0.0794	0.1397	0.4310	0.1760	0.0767	2.55 ± 0.01

Then, Rietveld refinements employing the FullProf subprogram were performed to calculate the crystal parameters of the *FeGaO-01* and *FeGaO-02* samples. The crystal structure was assumed as the inverse spinel crystal formation. The inverse spinel crystal has the same XRD pattern of the FeGa$_2$O$_4$ spinel crystal. Under the assumption of the inverse spinel crystal, the crystal structure of FeGa$_2$O$_4$ was taken into account as a cubic and the cations, Fe^{2+} and Ga^{3+}, were placed both the tetrahedral (0, 0, 0) and octahedral ($\frac{5}{8}, \frac{5}{8}, \frac{5}{8}$) sites. The space group of samples was assumed to be the F4d$_1$23 (No.210) space group, according to the fit of the MATCH subprogram. Then obtained parameters from the MATCH program were used in the FullProf program for structural analyses. The FullProf calculations are illustrated in Figure 3 and the data are shown in Table 2.

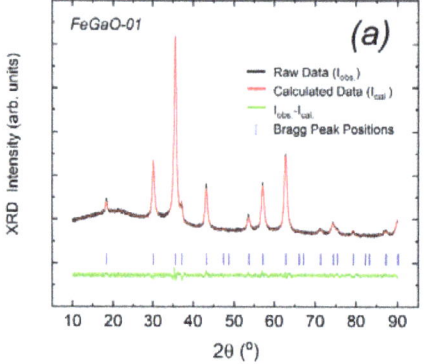
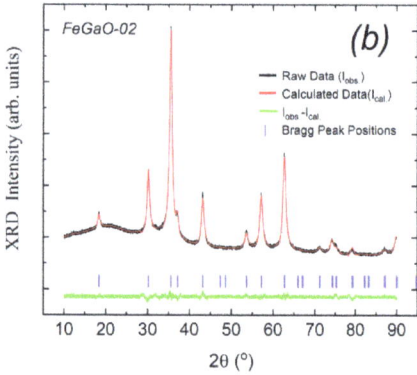

Figure 3. Rietveld refinements fits for the synthesized samples of (**a**) *FeGaO-01* and (**b**) *FeGaO-02*.

The bandgap values of each spectrum were analyzed from the reflectance spectra and each spectrum was modified by Tauc's Relation, as set out in Equation (5) [33]:

$$(\alpha h v)^n = A(h v - E g) \quad (5)$$

where Eg, h, v, α, A, and n represent the bandgap energy, Planck constant, frequency of light, absorption co-efficient, constant, and a constant for the direct band gap, 2, respectively. The fits are shown in Figure 4a,b, in which the horizontal axes are in the energy unit calculated by Equation (6):

$$E = h v = \frac{hc}{\lambda} \quad (6)$$

where E, λ, and c indicate the energy, wavelength, and speed of the light, respectively. The bandgap values of *FeGaO-01* and *FeGaO-02* were obtained as 2.39 ± 0.02 eV and 2.55 ± 0.01 eV. The difference between the calculated band gap values were assigned the shallow defect states due to the formation of partial inverse spinel states, or to excess Fe atoms, which could have come from the miller vial during the milling processes [34].

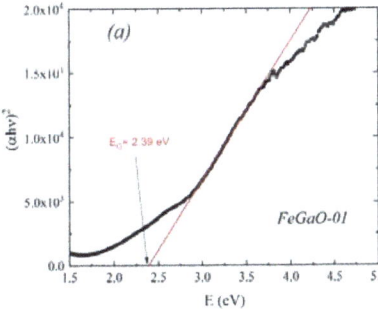 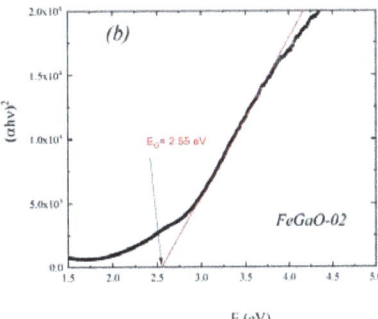

Figure 4. UV-vis spectra of the (**a**) *FeGaO-01* and (**b**) *FeGaO-02* samples.

As observed in the literature, excess atoms can stay in the crystal lattice as interstitial defect states [35]. These defect states create either donor or acceptor levels, which have electronic energy levels very close to band gap edges, possibly narrowing the band gap values [36].

4. Conclusions

The research was intended to optimize mechanochemical techniques to synthesize $FeGa_2O_4$ particles. The XRD patterns revealed that the $FeGa_2O_4$ particles were synthesized without any contamination elements or compounds. Replacement atoms in the tetrahedral and octahedral sites were analyzed by employing Reitveld refinement. Calculations were performed by assuming that the crystal configurations of the particles were formed as full or partial inverse spinel crystal structures. The bandgap values of $FeGa_2O_4$ were calculated as 2.39 ± 0.02 eV and 2.55 ± 0.01 eV. Measuring the two different bandgap values indicated the formation of shallow energy levels originating from defect energy states, which were close to either conduction or valance energy levels. Furthermore, the 84 h of milling caused the formation of a 10 times higher microstrain, according to 36 h milled particles. The high microstrain value broadened the peak values, and thus the result of the high microstrain was the decreased band gap value of the particles.

Author Contributions: Conceptualization, M.M.C.; methodology, M.M.C.; formal analysis, M.M.C. and Y.A.; investigation, M.M.C. and Y.A.; data curation, M.M.C. and Y.A.; writing—original draft preparation, M.M.C.; writing—review and editing, M.M.C. and S.K.; supervision, M.M.C.; project administration, M.M.C.; funding acquisition, M.M.C. All authors have read and agreed to the published version of the manuscript.

Funding: This research was funded by [the Scientific and Technological Research Council of Turkey (TÜBİTAK)] grant number [118F373]. This research was also funded by [the Scientific Research Projects Coordination Unit of Istanbul University], grant number [FYL-2021-38050].

Institutional Review Board Statement: Not applicable.

Informed Consent Statement: Not applicable.

Data Availability Statement: The data about space groups was acquired from the MATCH subprogram, and they have not given their permission for researchers to share their data. Data requests can be made to the company CRYSTAL IMPACT via this email: info@crystalimpact.de.

Conflicts of Interest: The authors declare that they have no conflict of interest.

References

1. Gao, H.; Liu, S.; Li, Y.; Conte, E.; Cao, Y. A Critical Review of Spinel Structured Iron Cobalt Oxides Based Materials for Electrochemical Energy Storage and Conversion. *Energies* **2017**, *10*, 1787. [CrossRef]
2. Li, Y.; Yuan, Z.; Meng, F. Spinel-Type Materials Used for Gas Sensing: A Review. *Sensors* **2020**, *20*, 5413. [CrossRef] [PubMed]
3. Šutka, A.; Gross, K.A. Spinel ferrite oxide semiconductor gas sensors. *Sens. Actuators B Chem.* **2016**, *222*, 95–105. [CrossRef]
4. Preethi, V.; Kanmani, S. Photocatalytic hydrogen production. *Mater. Sci. Semicond. Process.* **2013**, *16*, 561–575. [CrossRef]
5. Suresh, R.; Rajendran, S.; Kumar, P.S.; Vo, D.-V.N.; Cornejo-Ponce, L. Recent advancements of spinel ferrite based binary nanocomposite photocatalysts in wastewater treatment. *Chemosphere* **2021**, *274*, 129734. [CrossRef] [PubMed]
6. Grimes, R.W.; Anderson, A.B.; Heuer, A.H. Predictions of cation distributions in AB2O4 spinels from normalized ion energies. *J. Am. Chem. Soc.* **1989**, *111*, 1–7. [CrossRef]
7. Rafiq, M.A.; Javed, A.; Rasul, M.N.; Nadeem, M.; Iqbal, F.; Hussain, A. Structural, electronic, magnetic and optical properties of AB_2O_4 (A = Ge, Co and B = Ga, Co) spinel oxides. *Mater. Chem. Phys.* **2020**, *257*, 123794. [CrossRef]
8. Pinto, A. Magnetization and Anisotropy in Gallium Iron Oxide. *J. Appl. Phys.* **1966**, *37*, 4372–4376. [CrossRef]
9. Abrahams, S.C.; Reddy, J.M.; Bernstein, J.L. Crystal Structure of Piezoelectric Ferromagnetic Gallium Iron Oxide. *J. Chem. Phys.* **1965**, *42*, 3957–3968. [CrossRef]
10. Ghose, J.; Hallam, G.C.; Read, D.A. A magnetic study of $FeGa_2O_4$. *J. Phys. C Solid State Phys.* **1977**, *10*, 1051–1057. [CrossRef]
11. Huang, C.-C.; Su, C.-H.; Liao, M.-Y.; Yeh, C.-S. Magneto-optical $FeGa_2O_4$ nanoparticles as dual-modality high contrast efficacy T2 imaging and cathodoluminescent agents. *Phys. Chem. Chem. Phys.* **2009**, *11*, 6331–6334. [CrossRef] [PubMed]
12. Alldredge, L.M.B.; Chopdekar, R.; Nelson-Cheeseman, B.B.; Suzuki, Y. Spin-polarized conduction in oxide magnetic tunnel junctions with magnetic and nonmagnetic insulating barrier layers. *Appl. Phys. Lett.* **2006**, *89*, 182504. [CrossRef]
13. Myoung, B.R.; Han, S.K.; Kim, S.J.; Kim, C.S. The Magnetic Behaviors of Spin-Glass $FeGa_2O_4$ system. *IEEE Trans. Magn.* **2012**, *48*, 1567–1569. [CrossRef]
14. Sánchez, J.; Cortés-Hernández, D.A.; Escobedo-Bocardo, J.C.; Almanza-Robles, J.M.; Reyes-Rodríguez, P.Y.; Jasso-Terán, R.A.; Bartolo-Pérez, P.; De-León-Prado, L.E. Synthesis of $Mn_xGa_{1-x}Fe_2O_4$ magnetic nanoparticles by thermal decomposition method for medical diagnosis applications. *J. Magn. Magn. Mater.* **2017**, *427*, 272–275. [CrossRef]
15. He, Z.-H.; Gao, J.-F.; Kong, L.-B. Iron Gallium Oxide with High-Capacity and Super-Rate Performance as New Anode Materials for Li-Ion Capacitors. *Energy Fuels* **2021**, *35*, 8378–8386. [CrossRef]
16. Sánchez, J.; Prado, L.E.D.L.; Hernández, D.A.C. Magnetic Properties and Cytotoxicity of Ga-Mn Magnetic Ferrites Synthesized by the Citrate Sol-Gel Method. *Int. J. Chem. Mol. Eng.* **2017**, *11*, 590–595.
17. Paudel, T.R.; Zakutayev, A.; Lany, S.; D'Avezac, M.; Zunger, A. Doping Rules and Doping Prototypes in A_2BO_4 Spinel Oxides. *Adv. Funct. Mater.* **2011**, *21*, 4493–4501. [CrossRef]
18. Urso, C.; Barawi, M.; Gaspari, R.; Sirigu, G.; Kriegel, I.; Zavelani-Rossi, M.; Scotognella, F.; Manca, M.; Prato, M.; de Trizio, L.; et al. Colloidal Synthesis of Bipolar Off-Stoichiometric Gallium Iron Oxide Spinel-Type Nanocrystals with Near-IR Plasmon Resonance. *J. Am. Chem. Soc.* **2016**, *139*, 1198–1206. [CrossRef]
19. Myoung, B.R.; Lim, J.T.; Kim, C.S. Crystallograhic and Magnetic Properties of $Fe_{1-x}Ni_xGa_2O_4$. *J. Korean Phys. Soc.* **2017**, *70*, 85–88. [CrossRef]
20. Lyubutin, I.S.; Starchikov, S.S.; Gervits, N.E.; Lin, C.-R.; Tseng, Y.-T.; Shih, K.-Y.; Lee, J.-S.; Ogarkova, Y.L.; Korotkov, N.Y. Structural, magnetic and electronic properties of $Fe_{1+x}Ga_{2-x}O_4$ nanoparticles synthesized by the combustion method. *Phys. Chem. Chem. Phys.* **2016**, *18*, 22276–22285. [CrossRef]
21. Lyubutin, I.S.; Starchikov, S.S.; Gervits, N.E.; Lin, C.-R.; Tseng, Y.-T.; Shih, K.-Y.; Lee, J.-S.; Ogarkova, Y.L.; Baskakov, A.O.; Frolov, K.V. Magnetic Properties and Charge Transfer Transition Induced by Jahn–Teller Effect in $FeGa_2O_4$ Nanoparticles. *J. Phys. Chem. C* **2016**, *120*, 25596–25603. [CrossRef]
22. Xu, Z.; Yan, S.-C.; Shi, Z.; Yao, Y.-F.; Zhou, P.; Wang, H.-Y.; Zou, Z.-G. Adjusting the Crystallinity of Mesoporous Spinel $CoGa_2O_4$ for Efficient Water Oxidation. *ACS Appl. Mater. Interface* **2016**, *8*, 12887–12893. [CrossRef] [PubMed]
23. Sun, X.; Maeda, K.; Le Faucheur, M.; Teramura, K.; Domen, K. Preparation of $(Ga_{1-x}Zn_x)(N_{1-x}O_x)$ solid-solution from $ZnGa_2O_4$ and ZnO as a photo-catalyst for overall water splitting under visible light. *Appl. Catal. A Gen.* **2007**, *327*, 114–121. [CrossRef]
24. Zhang, Q.; Saito, F. A review on mechanochemical syntheses of functional materials. *Adv. Powder Technol.* **2012**, *23*, 523–531. [CrossRef]
25. Takacs, L. Self-sustaining reactions induced by ball milling. *Prog. Mater. Sci.* **2002**, *47*, 355–414. [CrossRef]
26. Janot, R.; Guérard, D. One-step synthesis of maghemite nanometric powders by ball-milling. *J. Alloys Compd.* **2002**, *333*, 302–307. [CrossRef]
27. Patterson, A.L. The Scherrer formula for X-ray particle size determination. *Phys. Rev.* **1939**, *56*, 978–982. [CrossRef]
28. Rabiei, M.; Palevicius, A.; Monshi, A.; Nasiri, S.; Vilkauskas, A.; Janusas, G. Comparing Methods for Calculating Nano Crystal Size of Natural Hydroxyapatite Using X-ray Diffraction. *Nanomaterials* **2020**, *10*, 1627. [CrossRef]
29. Mote, V.D.; Purushotham, Y.; Dole, B.N. Williamson-Hall analysis in estimation of lattice strain in nanometer-sized ZnO particles. *J. Theor. Appl. Phys.* **2012**, *6*, 6. [CrossRef]
30. Prabhu, Y.T.; Rao, K.V.; Kumar, V.S.S.; Kumari, B.S. X-ray Analysis by Williamson-Hall and Size-Strain Plot Methods of ZnO Nanoparticles with Fuel Variation. *World J. Nano Sci. Eng.* **2014**, *04*, 21–28. [CrossRef]

31. Zak, A.K.; Majid, W.H.A.; Abrishami, M.E.; Yousef, R. X-ray analysis of ZnO nanoparticles by WilliamsoneHall and sizeestrain plot methods. *Solid State Sci.* **2011**, *13*, 251–256.
32. Stukowski, A.; Markmann, J.; Weissmüller, J.; Albe, K. Atomistic origin of microstrain broadening in diffraction data of nanocrystalline solids. *Acta Mater.* **2009**, *57*, 1648–1654. [CrossRef]
33. Dolgonos, A.; Mason, T.O.; Poeppelmeier, K.R. Direct optical band gap measurement in polycrystalline semiconductors: A critical look at the Tauc method. *J. Solid State Chem.* **2016**, *240*, 43–48. [CrossRef]
34. Farooq, M.I.; Khan, M.S.; Yousaf, M.; Zhang, K.; Zou, B. Antiferromagnetic Magnetic Polaron Formation and Optical Properties of CVD-Grown Mn-Doped Zinc Stannate (ZTO). *ACS Appl. Electron. Mater.* **2020**, *2*, 1679–1688. [CrossRef]
35. Shigetomi, S.; Ikari, T. Optical properties of GaSe grown with an excess and a lack of Ga atoms. *J. Appl. Phys.* **2003**, *94*, 5399. [CrossRef]
36. Walsh, A.; Da Silva, J.L.F.; Wei, S.-H. Origins of band-gap renormalization in degenerately doped semiconductors. *Phys. Rev. B* **2008**, *78*, 075211. [CrossRef]

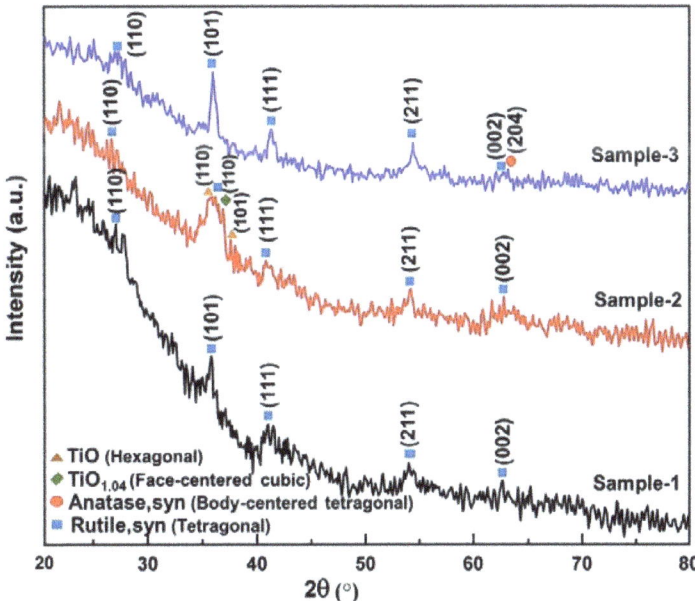

Figure 2. Thin-film X-ray diffraction patterns with different sputtering process parameters thermally oxidized at 450 °C.

Figure 3 shows the XRD patterns of samples with different sputtering process parameters annealed at 550 °C. A further increase in temperature to 550 °C resulted in increased diffraction intensity with an associated decrease in the XRD patterns' amorphous background, indicating improved Ti film crystallinity. Only a rutile crystallization phase was clearly observed in samples 1 and 3 at 550 °C. This indicates complete anatase transition into the rutile phase at 550 °C for samples 1 and 3. As is well known, anatase is a metastable phase that may be transformed into the rutile phase during heat treatment [18,28]. For Ti film thermal oxidation, Ti atoms absorb oxygen from the air into the films. The Ti and O atoms react to form TiO_2 when the oxidation temperature is 550 °C. However, there was a main rutile phase with a minor TiO phase in sample 2 oxidized at 550 °C. This result indicates that TiO_2 (as-deposited) as the bottom layer does not provide sufficient oxygen to the upper layer during thermal oxidation at 550 °C. Chung et al. reported that the formation of TiO was from the Ti thin film after thermal oxidation at 500–550 °C [29].

In the case of the rutile tetragonal crystal structure ($a = b \neq c$), the lattice constant is calculated using the following formula:

$$\frac{1}{d_{hkl}^2} = \frac{h^2 + k^2}{a} + \frac{l^2}{c^2}, \tag{3}$$

where d_{hkl} is the interplanar separation corresponding to Miller indices h, k, and l and a and c are the lattice constants. For all samples, the calculated structural parameters corresponding to peaks (110) and (101) are given in Table 1. When compared with standard rutile TiO_2 ($a = b = 4.5933$ Å and $c = 2.9592$ Å), a slight increase in the lattice constant a and a decrease in the lattice constant c for sample 2 when thermally oxidized at 450 °C were observed. This is due to the TiO and $TiO_{1.04}$ crystal peaks that represent an oxygen vacancy in a thin-film sample.

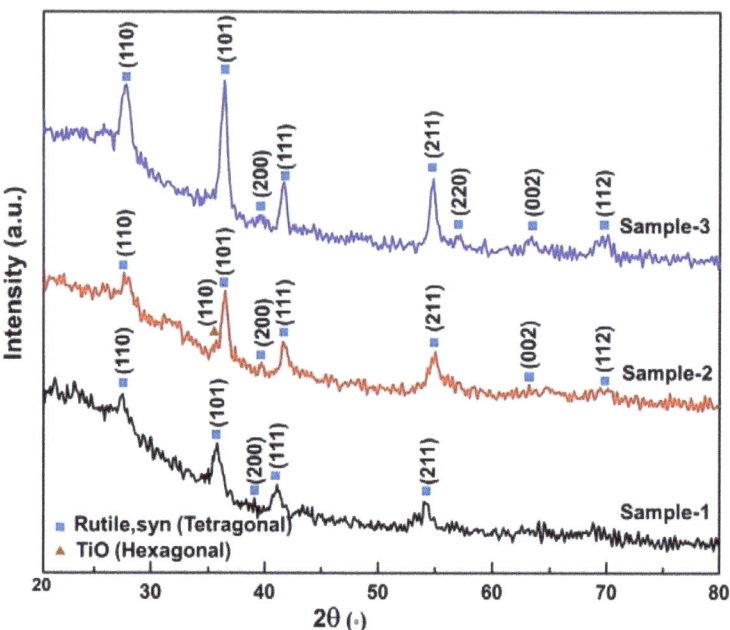

Figure 3. Thin-film X-ray diffraction patterns with different sputtering process parameters thermally oxidized at 550 °C.

Table 1. Optimized lattice parameters of the rutile phase in Ti-TiO$_2$ bilayer thin-film samples as a function of oxidation temperature.

Oxidation Temperature (°C)	Sample	a (Å)	c (Å)	Unit Cell Volume (Å^3)
	Sample 1	4.5501	2.9685	61.4581
450	Sample 2	4.6147	2.9400	62.6086
	Sample 3	4.5990	2.9649	62.7100
	Sample 1	4.5816	2.9942	62.8510
550	Sample 2	4.5829	2.9708	62.3964
	Sample 3	4.6145	2.9734	63.3147

High-resolution TEM (HR-TEM) was used to analyze TiO$_2$ thin films to determine whether they were crystalline or amorphous. The TEM specimen microstructures (samples 1, 2, 3) thermally oxidized at 450 °C are shown in Figures 4–6, respectively. Figure 4 shows typical bright- and dark-field HR-TEM images of sample 1 thermally oxidized at 450 °C. A film thickness of 120 nm was observed, as shown in Figure 4a. From the dark-field microstructure view, an interface was apparent in the film, separating the crystalline and amorphous zones, as shown in Figure 4b. The oxidation zone in the Ti films was about 57 nm less than the 63 nm amorphous zones. The Ti films thermally oxidized at 450 °C with crystalline structures are exhibited in the upper zone in Figure 4c,d. According to HR-TEM micrograph analysis, the crystalline grains belonged to the rutile crystalline phase (111). Figure 4e clearly shows that the bottom-zone film microstructure is amorphous. However, the film near the surface layer has obviously crystallization. Oxygen diffuses into the internal film zone from the surface by thermal oxidation at 450 °C.

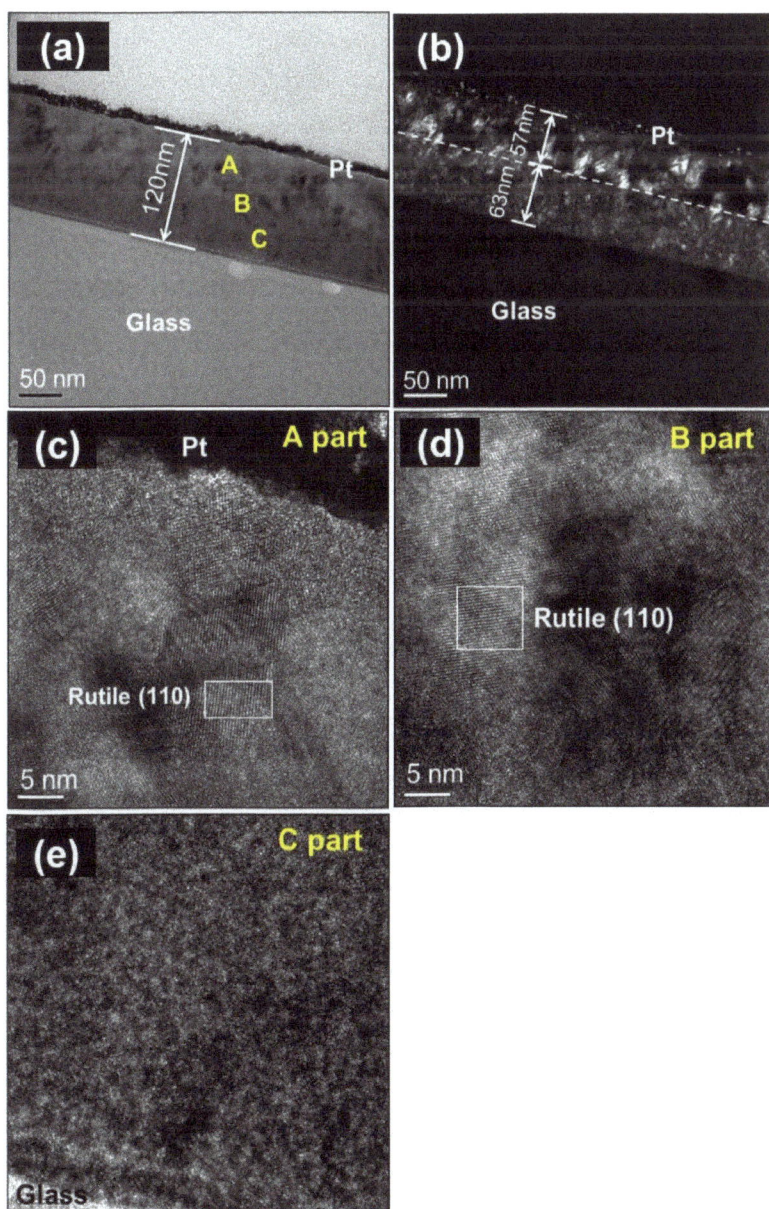

Figure 4. TEM micrographs of a sample 1 thin film thermally oxidized at 450 °C: (**a**) bright field, (**b**) dark field, (**c**) HR-TEM image showing point A, (**d**) HR-TEM image showing point B, and (**e**) HR-TEM image showing point C.

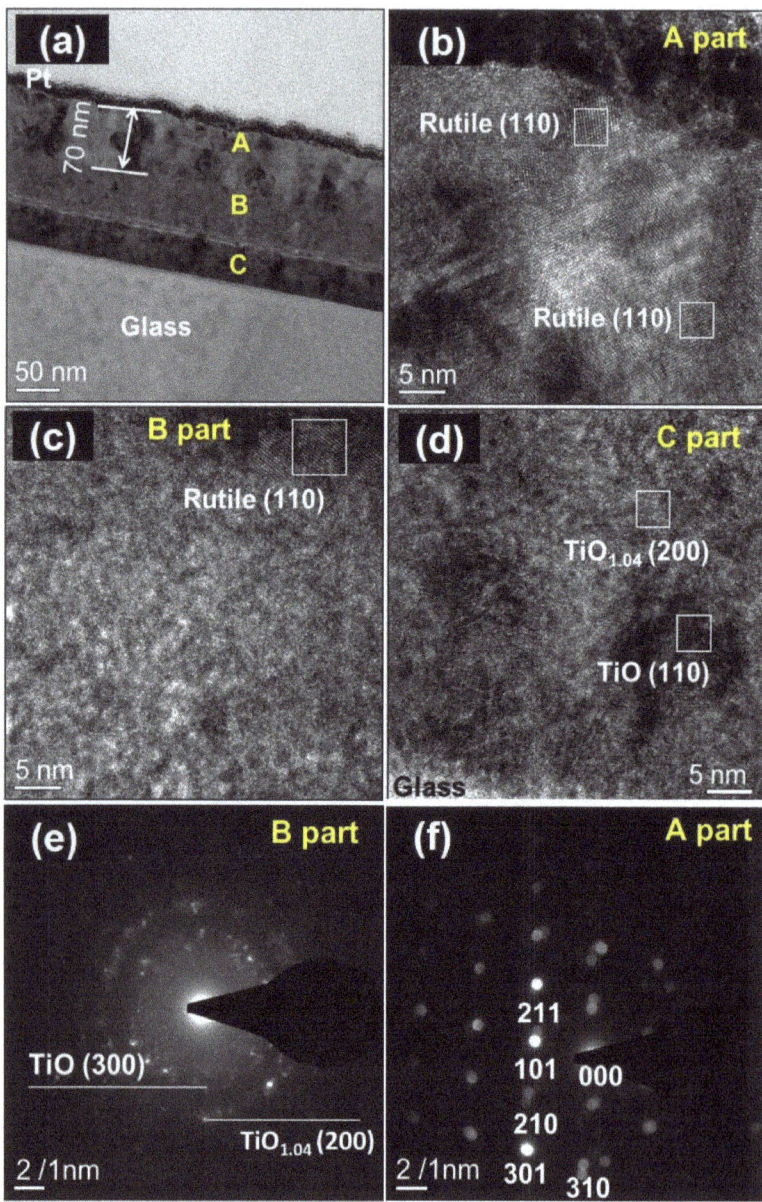

Figure 5. TEM micrographs of a sample 2 thin-film thermally oxidized at 450 °C: (**a**) plane view of the film, (**b**) HR-TEM image showing point A, (**c**) HR-TEM image showing point B, (**d**) HR-TEM image showing point C, (**e**) NBED showing point B, and (**f**) NBED showing point A.

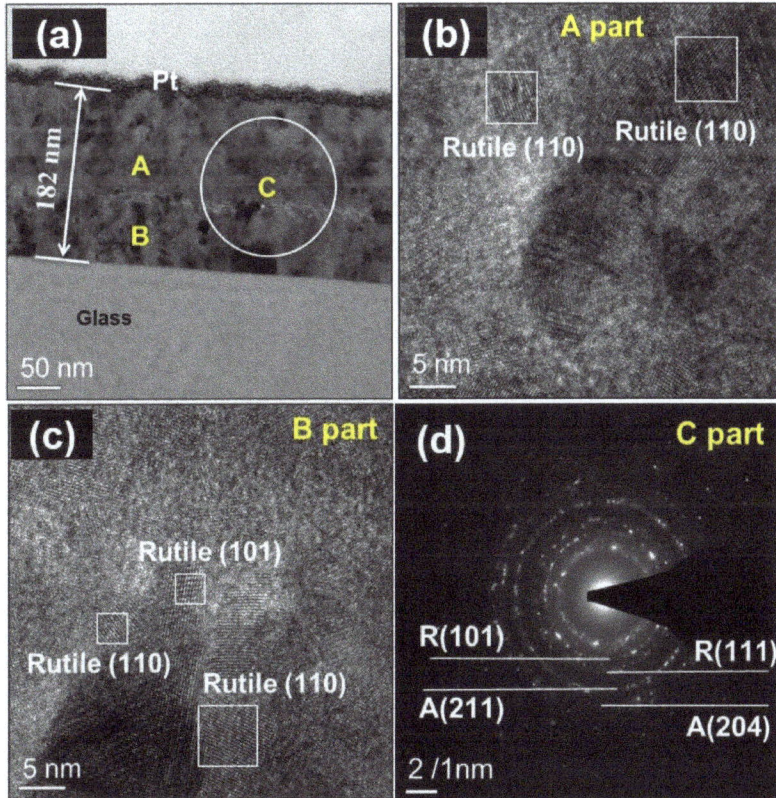

Figure 6. TEM micrographs of a sample 3 thin-film thermally oxidized at 450 °C; (**a**) plane view of the film, (**b**) HR-TEM image showing point A, (**c**) HR-TEM image showing point B, and (**d**) NBED showing point C.

Figure 5 shows sample 2 HR-TEM micrographs and SAED patterns from thermal oxidation at 450 °C for 4 h. The microstructure shows the same interface in the film. It can be divided into two zones. The oxidation thickness in Ti films was about 70 nm at 450 °C for 4 h thermal oxidation, as shown in Figure 5a, which was higher than that for sample 1 (57 nm). According to HR-TEM micrograph analysis, the upper-zone (apart) crystalline grains belong to the rutile crystalline phase (111), as shown in Figure 5b. Two kinds of phase structures coexisted in the interface zone (B part), amorphous and TiO_2 grains, as shown in Figure 5c. It clearly shows that the film microstructure is amorphous on the Ti film bottom zone. The TiO_2 film bottom layer C part was also analyzed, as shown in Figure 5d. The $TiO_{1.04}$ (200) and TiO (110) microcrystalline phases coexisted in bottom-layer films. This is attributed to the TiO_x film bottom layer deposited onto glass without annealing, which does not readily form the TiO_2 crystalline phase stoichiometry. Sekhar et al. reported that TiO_2 films deposited using DC reactive magnetron sputtering are amorphous in a continuous gas flow if no bias voltage or heating is applied to the substrate [30,31]. Electron diffraction pattern analysis was carried out to identify crystalline grains, as shown in Figure 5e,f. The SAED pattern displayed a rutile phase in the upper-zone A part (Figure 5f). However, the SAED pattern showed $TiO_{1.04}$ (200) and TiO (300) planes on the interface-zone B part (Figure 5e). This observation is consistent with the XRD results in the above section. TiO existed in the bottom layer, implying a lack of oxygen content in the film. The oxygen partial pressure influences the crystal formation of Ti films during sputtering [32].

Figure 6 shows HR-TEM micrographs and SAED patterns of sample 3 thermally oxidized at 450 °C. It was mentioned in the above section that sample 3 was TiO_2 films deposited on glass substrates as the bottom layer, and it was annealed at 550 °C for 3 h first, and then pure titanium film was deposited on the bottom layer (TiO_2 films). In the dark-field view, there was no amorphous zone observed in the films, as shown in Figure 6a. This shows that the film has good crystallinity in all zones, as shown in Figure 6b,c. According to SAED pattern analysis, the crystalline grains belong to the rutile and anatase phases in the films, as shown in Figure 6d. The related literature reported that in the reactive sputtering process, the rutile-and-anatase-mixture-phase TiO_2 growth is induced, and the anatase phase increases with the increase in oxygen in the sputter chamber [31,32].

As is well known, anatase phase formation requires sufficient oxygen. Comparing sample 1 to sample 3, the oxygen diffusion from the bottom titanium oxide layer influences the microstructure and phase evolution of Ti films during thermal oxidation. For example, mixture phases (rutile and anatase) with good crystallinity in oxidized Ti films can be obtained when the bottom layer is rutile TiO_2 films. Figure 7 shows the amorphous layer thickness of the Ti films on different bottom layers as a function of the oxidation temperature. The amorphous layer thickness was analyzed using TEM. The results showed that the amorphous layer thickness depends strongly on the bottom layer's oxygen concentration and the oxidation temperature for all samples. The amorphous zone decreased significantly with increasing thermal oxidation temperature. However, a film without an amorphous structure (fully crystalline) was found at 450 °C for sample 3 and at 550 °C for samples 1 and 2. This explains how annealed titanium oxide films as the bottom layer lead to Ti films becoming TiO_2 crystalline phases after 450 °C thermal oxidation.

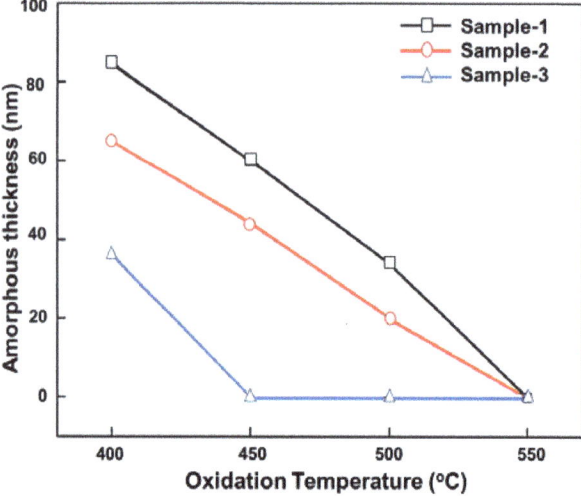

Figure 7. Amorphous thickness of Ti films measured as a function of oxidation temperature.

XPS analysis was performed to observe the phase evolution between different sample conditions of Ti-TiO_2 bilayer thin films and the oxidation states of the constituent elements present in the samples. Figure 8 displays the spectra of Ti-TiO_2 bilayer thin films on glass with different oxidation temperatures. Figure 8a shows the high-resolution XPS spectrum of the Ti 2p peak for samples with different bottom layers at 450 °C. The binding energy curves of Ti 2p with two firm peaks were observed at ~457 and ~462 eV, corresponding to core levels of Ti $2p_{3/2}$ and Ti $2p_{1/2}$, respectively, which is in good agreement with the recent literature [33,34]. The energy binding differences between the two peaks are dependent on the Ti atoms' chemical state and the value between these spin-split components [35].

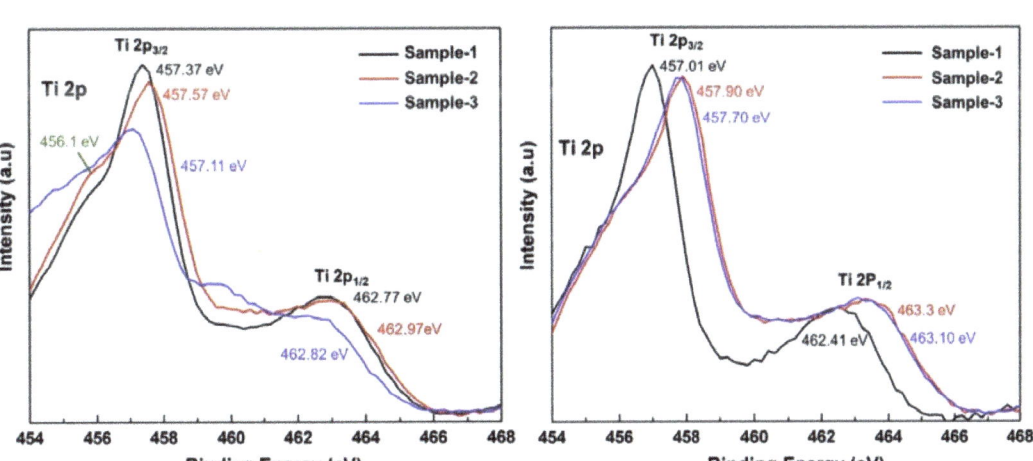

Figure 8. XPS curve of the internal area of the Ti-TiO$_2$ bilayer thin film at different oxidation temperatures: binding energy curves of Ti 2p, employing a Shirley-type background oxidized at (**a**) 450 °C and (**b**) 550 °C.

It was observed from the XPS spectra in Figure 8b that increasing the oxidation temperature up to 550 °C shows the binding energy peaks of Ti 2p at single-layer films shifting toward a higher binding energy by ~0.7 eV for both bilayer thin films (samples 2 and 3). Fakhouri et al. [36] proposed that increasing the thin-film layer results in the Ti 2p peak shifting toward lower or high binding energy. This phenomenon may be due to a partial reduction in the titanium cations related to the formation of oxygen vacancies in the lattice or a change in stacked electronic affinity at Ti-TiO$_2$ thin films [36]. According to the literature [37], the chemical states for all thin-film samples (thermally oxidized at 450 and 550 °C) correspond to Ti^{3+} due to the formation of Ti–O bonds at the binding energy of Ti 2p$_{3/2}$ of around ~457 eV, characteristic of rutile titanium dioxide that is similar to TiO$_2$. However, the small Ti^{2+} peaks (456.1 eV) in sample 2 indicated the presence of TiO. The presence of TiO is consistent with the HR-TEM result in Figure 5.

The compositions of Ti-TiO$_2$ bilayer thin films at different thermal oxidation temperatures were identified using XPS. The relative concentrations of titanium and oxygen were analyzed in the Ti-TiO$_2$ bilayer thin films. To understand the difference in oxygen concentration between the surface and the internal zone of the upper Ti oxidized films, the film compositions were analyzed at different positions. The average values of each element are listed in Table 2. The oxygen concentration in the internal zone was lower than that on the surface. For bilayer samples oxidized at 450 °C, the internal and surface zone's oxygen content was <59% and >69%, respectively. According to XPS analysis results, it is believed that the upper Ti films' internal zone is not fully oxidized. This means that many oxygen vacancies exist in the films. When the oxidation temperature was enhanced from 450 to 550 °C, the internal zone's oxygen content increased up to ~60%. The results indicate that the oxygen content in the upper Ti films increases by increasing the oxidation temperature.

Table 2. Chemical composition of the Ti-TiO$_2$ bilayer thin-film surface and internal area with different samples oxidized for 3 h.

Oxidation Temperature (°C)	Element	Atomic Concentration (%)					
		Sample 1		Sample 2		Sample 3	
		a	b	a	b	a	b
450	Ti	28.34	39.10	30.47	41.22	30.70	42.60
	O	71.66	60.90	69.53	58.78	69.30	57.40
550	Ti	15.46	35.56	29.91	40.01	23.46	39.96
	O	84.54	64.44	70.09	59.99	76.54	60.04

a Surface area; b internal area.

UV–VIS spectroscopy was conducted to measure the optical transmission of the Ti-TiO$_2$ bilayer thin films with different samples thermally oxidized at 450 and 550 °C, as shown in Figure 9. The Ti-TiO$_2$ bilayer thin-film transmittance spectra curves indicated that the transmittance spectra gradually shift to a short wavelength as the amorphous thickness in the film increases.

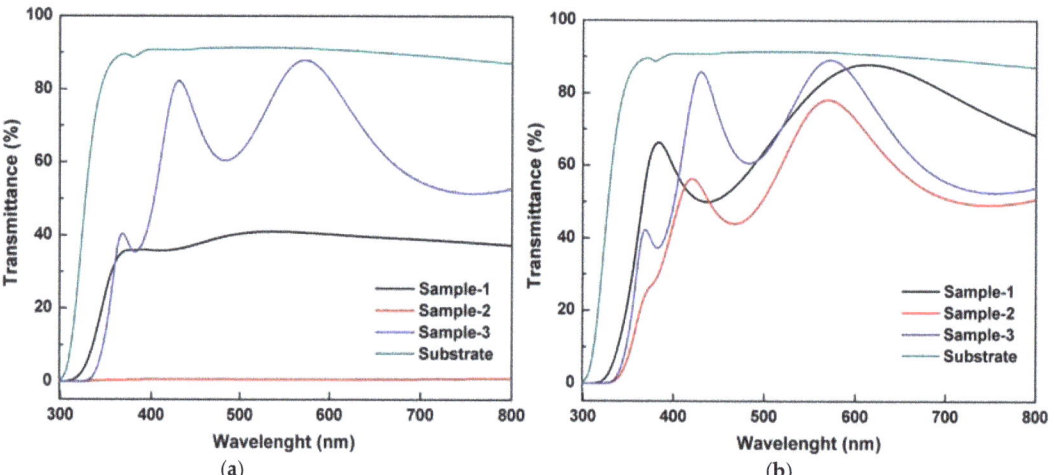

Figure 9. UV–VIS transmittance spectra of a thin film with different samples oxidized at (**a**) 450 °C and (**b**) 550 °C.

The optical transmission spectra of sample 3 oxidized at 450 °C were exceptionally clear and more than 86% in the visible region. In contrast, sample 1 thermally oxidized at 450 °C had lower transmittance (~41%), while sample 2 had almost no light transmission (0%), as seen in Figure 9a. According to the sample 1 microstructure analysis (Figure 4), oxygen diffusion into the inner film zone slowly occurred at 450 °C, clearly showing that the bottom-zone film microstructure is amorphous. Films with a larger amorphous zone can cause more reflections and light interference, resulting in weak light transmission [38].

The poorest transmittance, especially in the visible region, occurred in sample 2 when oxidized at 450 °C, which is attributed to high absorption in the metallic layer [28], suggesting a sufficient amount of oxygen vacancies that significantly absorbed the incident light [29]. This finding seems consistent with the XRD patterns (Figure 2) and HR-TEM image in Figure 5e, indicating that sample 2 has TiO$_{1.04}$ (200) and TiO (110) microcrystalline phases that coexist in bottom-layer films. TiO existed in the bottom layer, implying the lack of oxygen content in the film. These phenomena resulted in poor light transmission in sample 2, although it had a lower amorphous thickness than sample 1.

Optical transmittance can be significantly improved while increasing the oxidation temperature up to 550 °C. Due to this phenomenon, titanium metal changes to titanium dioxide after thermal oxidization and increases the oxidation layer of the bilayer thin film. Figure 9b displays bilayer thin films on samples 2 and 3, which showed a significant increase in transmission to ~78% and ~88%, respectively. The transmittance result from sample 3 was higher than the Ti single-layer film thermally oxidized at 600 °C reported by some authors [20,39]. The increasing oxidation temperature has been documented to provide adequate energy to diffuse into the film, improving the TiO_2 crystalline strength [40]. However, the transmittance of sample 2 was lower than other samples. It corresponded to TiO (110) instead of the rutile crystallization, as shown in the XRD pattern in Figure 3. As mentioned before, TiO exhibited higher reflective properties due to inadequate oxygen in films. Thin-film transparency improved by increasing the thermal oxidation temperature, which probably removed the residual stress, quantum confinement, and structure order improvement [25], and decreasing the amorphous thickness in the Ti-TiO_2 films.

The Ti-TiO_2 bilayer thin films' photocatalytic activity was determined by the degradation of 10 mg/L of MB solution under UV-C light irradiation, as shown in Figure 10. It can be observed that all thin-film samples that were thermally oxidized at 450 and 550 °C exhibited quick adsorption in the dark. This could be attributed to the high specific surface area and small particle size [41]. Figure 10a shows MB degradation (%) of a thin film with different samples oxidized at 450 °C. It was found that all samples provided high photocatalytic degradation performance with a range of 43–63% compared to the substrate. However, increasing the thermal oxidation temperature to 550 °C decreased the photocatalytic activity of samples 2 and 3 (Figure 10b).

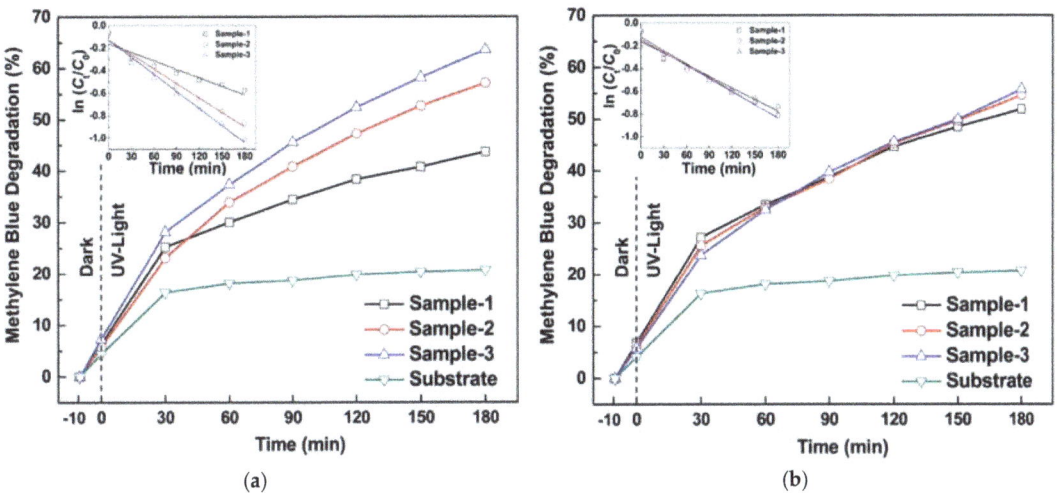

Figure 10. Photocatalytic activity of thin-film samples as the function of the irradiation time in MB solution. Inset: first-order reaction rate of thin-film samples with MB. MB degradation of thin-film samples thermally oxidized at (**a**) 450 °C and (**b**) 550 °C.

According to first-order reaction kinetics, in Figure 10 (inset), the time-dependent $\ln(C_t/C_0)$ terms are illustrated for all thin-film samples. The calculated reaction rate constants (k) of the thin films thermally oxidized at 450 °C with different sample preparations (samples 1, 2, and 3) were 2.52×10^{-3}, 4.29×10^{-3}, and 5.07×10^{-3} min^{-1}, respectively (Figure 10a inset), showing a higher rate of degradation for sample 3. The reaction constants (k) gradually increased with a decrease in the amorphous thickness. The amorphous area depended strongly on the bottom-layer annealing temperature and oxygen concentration for all samples. However, after increasing the oxidation temperature to 550 °C, the rate constants (k) of all thin-film samples were stable around 3×10^{-3} min^{-1} (Figure 10b inset). This is attributed to the crystalline phase changing from mixture phases (rutile and anatase) to a single phase (rutile) at 550 °C, as seen in Figures 2 and 3. It is well known that the combination of TiO$_2$ rutile and anatase phases exhibits higher photocatalytic activity in MB solution decomposition than the pure TiO$_2$ phase due to the electrons' movement from rutile to anatase TiO$_2$ during photoexcitation. This prevents anatase charge recombination, leading to more efficient photogenerated electron–hole pair separation and high photocatalytic performance [42,43]. Kalarivalappil et al. [44] reported that the highest photocatalytic activity is obtained in thin films with an optimum ratio of anatase and rutile phases. However, higher rutile content lessens their photocatalytic activity due to the photogenerated holes and the electrons' higher recombination rate in the rutile phase.

Figure 11 illustrates the photocatalytic performance of Ti-TiO$_2$ thin films thermally oxidized at different temperatures. It can be clearly seen that both bilayer thin films (samples 2 and 3) had higher photocatalytic performance than the single-layer film (sample 1) due to the lower amorphous thickness in bilayer thin films, as shown in Figure 7. This finding is in agreement with previous research [45].

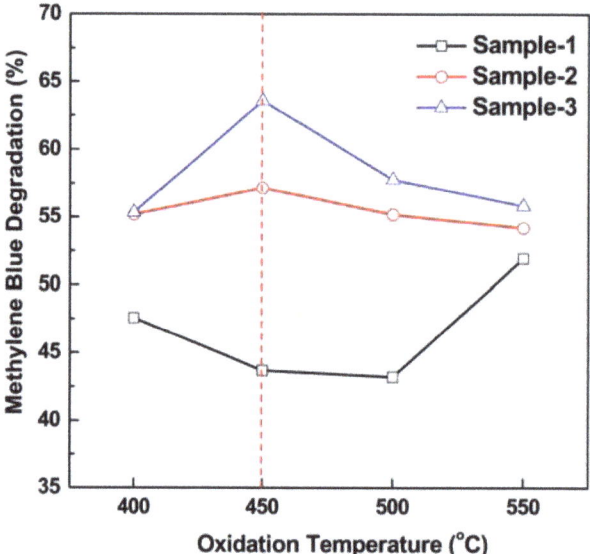

Figure 11. Photocatalytic activity of TiO$_2$ thin-film samples at different oxidation temperatures irradiated at 180 min.

It was found that the oxidation temperature affects the sample photocatalytic activity. As the thermal oxidation temperature rose from 400 to 450 °C, both bilayer thin films' degradation rate increased, as seen in samples 2 (57%) and 3 (63%). However, after increasing the thermal oxidation temperature to 550 °C, the bilayer thin-film degradation rate steeply dropped. In addition, sample 2 thermally oxidized at 450 °C showed good photocatalytic activity to degrade MB solution instead of lower optical transmittance. The existence of TiO$_{1.04}$ (200), TiO (110), and rutile phases in sample 2 contributed to poor transmittance results, while good photocatalytic activity (Figure 7) was due to a higher number of oxygen vacancy defects in the films, contributing to increasing photocatalytic activity. As is well known, oxygen vacancies play a vital role in semiconductor photocatalysis. Therefore, inducing large effects on the electronic, photonic, and photocatalytic properties of TiO$_2$, oxygen deficiencies are critical features [46–48].

Elahifard et al. [49] developed TiO$_2$ nano-materials with controllable VO contents (0–2.18%) using an in situ solid-state chemical reduction strategy. The work showed that the bandgap of the resultant rutile TiO$_2$ reduced from 3.0 to 2.56 eV, indicating enhanced visible light absorption. Furthermore, semiconducting photocatalysts are mainly dependent upon the separation of photogenerated electron–hole pairs and transfer of electrons from the photocatalyst into the organic pollutants through the oxygen vacancy defects on the photocatalyst surface [50,51]. Stojadinović et al. [52] reported different photocatalytic activities of Tb^{3+}-doped TiO_2 coatings, in which a thicker TiO_2 coating oxidation layer has a higher number of oxygen vacancy defects, with the oxide contributing to the increasing photocatalytic activity.

These results indicated that oxygen vacancies are crucial parameters for obtaining high photocatalytic performance. The annealed bottom layer can provide high oxygen diffusion to upper Ti thin films and create a higher number of oxygen vacancy defects in the films, contributing to the increasing photocatalytic activity. Furthermore, an optimum oxidation temperature (450 °C) from TiO_2 rutile and anatase phases exhibits higher photocatalytic activity in MB solution degradation than the pure rutile TiO_2 phase. Thus, the study demonstrates that this bilayer modification strategy promotes the oxygen-induced TiO_2 bilayer thin-film bottom layer that can increase thin films' transmittance and photocatalytic activity. Furthermore, bilayer thin films could have the potential as photocatalytic coating material, including self-cleaning, self-sterilizing (antimicrobial), and water purification.

4. Conclusions

The bottom-layer oxygen dependency of thin-film samples was investigated in detail by using three types of samples (a single layer of Ti, a bilayer of Ti-TiO_2 as-deposited, and a bilayer of Ti-TiO_2 annealed thin films). It was found that the film's phase structure evolution after thermal oxidation was significantly affected by oxygen diffusion from the bottom to the upper layer. Upon further increasing the thermal oxidation temperature, all samples were observed to have a single rutile phase and a notable decline in the amorphous zone in bilayer thin films. The oxygen vacancy is a crucial parameter for obtaining high photocatalytic performance. The annealed bottom layer can provide high oxygen diffusion to upper Ti thin films and create a higher number of oxygen vacancy defects in the films, contributing to increasing photocatalytic activity. This study demonstrates that the bilayer modification strategy promotes the oxygen-induced bottom layer of TiO_2 bilayer thin films.

Author Contributions: Methodology and data curation, P.-Y.L.; writing—original draft preparation, writing—review and editing and validation, E.W.; visualization and formal Analysis, T.-C.L.; conceptualization and methodology C.-T.C.; Software and investigation, F.-Y.X.; formal analysis and resources, Y.-T.T.; supervision and funding acquisition, Y.-C.L. All authors have read and agreed to the published version of the manuscript.

Funding: This research received no external funding.

Institutional Review Board Statement: Not applicable.

Informed Consent Statement: Not applicable.

Data Availability Statement: The data presented in this study are available on request from the corresponding author.

Acknowledgments: The author wishes to thank the Department of Materials Engineering, National Pingtung University of Science and Technology, for support.

Conflicts of Interest: The authors declare no conflict of interest.

References

1. Chen, L.; Graham, M.E.; Li, G.; Gray, K.A. Fabricating highly active mixed phase TiO_2 photocatalysts by reactive DC magnetron sputter deposition. *Thin Solid Film* **2006**, *515*, 1176–1181. [CrossRef]
2. Yu, J.; Zhao, X.; Zhao, Q. Photocatalytic activity of nanometer TiO_2 thin films prepared by the Sol–Gel Method. *Mater. Chem. Phys.* **2001**, *69*, 25–29. [CrossRef]
3. Francioso, L.; Presicce, D.S.; Siciliano, P.; Ficarella, A. Combustion conditions discrimination properties of Pt-doped TiO_2 thin film oxygen sensor. *Sens. Actuators B Chem.* **2007**, *123*, 516–521. [CrossRef]

4. Weihao, L.; Shengnan, C.; Jianxun, W.; Shaohui, J.; Chenlu, S.; Yong, L.; Gaorong, H. Study on structural, optical and hydrophilic properties of FTO/TiO$_2$ tandem thin film prepared by aerosol-assisted chemical vapor deposition method. *Surf. Coat. Technol.* **2019**, *358*, 715–720. [CrossRef]
5. Greczynski, G.; Petrov, I.; Greene, J.E.; Hultman, L. Paradigm shift in thin-film growth by magnetron sputtering: From gas-ion to metal-ion irradiation of the growing film. *J. Vac. Sci. Technol. A Vac. Surf. Film* **2019**, *37*, 60801. [CrossRef]
6. Jain, R.; Shrivastava, M. Photocatalytic removal of hazardous dye cyanosine from industrial waste using titanium dioxide. *J. Hazard. Mater.* **2008**, *152*, 216–220. [CrossRef]
7. Yao, K.S.; Wang, D.Y.; Chang, C.Y.; Weng, K.W.; Yang, L.Y.; Lee, S.J.; Cheng, T.C.; Hwang, C.C. Photocatalytic disinfection of phytopathogenic bacteria by dye-sensitized TiO$_2$ thin film activated by visible light. *Surf. Coat. Technol.* **2007**, *202*, 1329–1332. [CrossRef]
8. Ferrec, A.; Keraudy, J.; Jacq, S.; Schuster, F.; Jouan, P.-Y.; Djouadi, M.A. Correlation between mass-spectrometer measurements and thin film characteristics using DcMS and HiPIMS discharges. *Surf. Coat. Technol.* **2014**, *250*, 52–56. [CrossRef]
9. Del Giudice, L.; Adjam, S.; La Grange, D.; Banakh, O.; Karimi, A.; Sanjinés, R. NbTiN thin films deposited by hybrid HiPIMS/DC magnetron co-sputtering. *Surf. Coat. Technol.* **2016**, *295*, 99–106. [CrossRef]
10. Greczynski, G.; Mráz, S.; Hans, M.; Primetzhofer, D.; Lu, J.; Hultman, L.; Schneider, J.M. Unprecedented Al supersaturation in single-phase rock salt structure VAlN films by Al+ subplantation. *J. Appl. Phys.* **2017**, *121*, 171907. [CrossRef]
11. Shimizu, T.; Komiya, H.; Teranishi, Y.; Morikawa, K.; Nagasaka, H.; Yang, M. Pressure dependence of (Ti, Al) N film growth on inner walls of small holes in high-power impulse magnetron sputtering. *Thin Solid Film.* **2017**, *624*, 189–196. [CrossRef]
12. Zauner, L.; Ertelthaler, P.; Wojcik, T.; Bolvardi, H.; Kolozsvári, S.; Mayrhofer, P.H.; Riedl, H. Reactive HiPIMS Deposition of Ti-Al-N: Influence of the deposition parameters on the cubic to hexagonal phase transition. *Surf. Coat. Technol.* **2020**, *382*, 125007. [CrossRef]
13. Lou, B.-S.; Yang, Y.-C.; Qiu, Y.-X.; Diyatmika, W.; Lee, J.-W. Hybrid high power impulse and radio frequency magnetron sputtering system for TiCrSiN thin film depositions: Plasma characteristics and film properties. *Surf. Coat. Technol.* **2018**, *350*, 762–772. [CrossRef]
14. Fan, Q.; Liang, Y.; Wu, Z.; Liu, Y.; Wang, T. Microstructure and properties of CrAlSiN coatings deposited by HiPIMS and direct-current magnetron sputtering. *Coatings* **2019**, *9*, 512. [CrossRef]
15. Greczynski, G.; Lu, J.; Jensen, J.; Bolz, S.; Kölker, W.; Schiffers, C.; Lemmer, O.; Greene, J.E.; Hultman, L. A Review of metal-ion-flux-controlled growth of metastable TiAlN by HIPIMS/DCMS co-sputtering. *Surf. Coat. Technol.* **2014**, *257*, 15–25. [CrossRef]
16. Greczynski, G.; Lu, J.; Jensen, J.; Petrov, I.; Greene, J.E.; Bolz, S.; Kölker, W.; Schiffers, C.; Lemmer, O.; Hultman, L. Strain-free, single-phase metastable Ti$_{0.38}$Al$_{0.62}$N alloys with high hardness: Metal-ion energy vs. momentum effects during film growth by hybrid high-power pulsed/Dc magnetron cosputtering. *Thin Solid Film.* **2014**, *556*, 87–98. [CrossRef]
17. Sartale, S.D.; Ansari, A.A.; Rezvani, S.J. Influence of Ti film thickness and oxidation temperature on TiO$_2$ thin film formation via thermal oxidation of sputtered Ti film. *Mater. Sci. Semicond. Process.* **2013**, *16*, 2005–2012. [CrossRef]
18. Zhou, B.; Jiang, X.; Liu, Z.; Shen, R.; Rogachev, A.V. Preparation and characterization of TiO$_2$ thin film by thermal oxidation of sputtered Ti film. *Mater. Sci. Semicond. Process.* **2013**, *16*, 513–519. [CrossRef]
19. Munguti, L.; Dejene, F. Influence of annealing temperature on structural, optical and photocatalytic properties of ZnO–TiO$_2$ composites for application in dye removal in water. *Nano-Struct. Nano-Objects* **2020**, *24*, 100594. [CrossRef]
20. Astinchap, B.; Moradian, R.; Gholami, K. Effect of sputtering power on optical properties of prepared TiO$_2$ thin films by thermal oxidation of sputtered Ti layers. *Mater. Sci. Semicond. Process.* **2017**, *63*, 169–175. [CrossRef]
21. Rothschild, A.; Edelman, F.; Komem, Y.; Cosandey, F. Sensing behavior of TiO$_2$ thin films exposed to air at low temperatures. *Sens. Actuators B Chem.* **2000**, *67*, 282–289. [CrossRef]
22. Vaquila, I.; Passeggi, M.C.G.; Ferrón, J. Oxidation process in titanium thin films. *Phys. Rev. B* **1997**, *55*, 13925. [CrossRef]
23. Martin, M.; Mader, W.; Fromm, E. Oxidation of iron, aluminium and titanium films in the temperature range 50–200 °C. *Thin Solid Film* **1994**, *250*, 61–66. [CrossRef]
24. Aniołek, K. The influence of thermal oxidation parameters on the growth of oxide layers on titanium. *Vacuum* **2017**, *144*, 94–100. [CrossRef]
25. Ting, C.-C.; Chen, S.-Y.; Liu, D.-M. Structural evolution and optical properties of TiO$_2$ thin films prepared by thermal oxidation of sputtered Ti films. *J. Appl. Phys.* **2000**, *88*, 4628–4633. [CrossRef]
26. Sarvadii, S.Y.; Gatin, A.K.; Kharitonov, V.A.; Dokhlikova, N.V.; Ozerin, S.A.; Grishin, M.V.; Shub, B.R. Oxidation of thin titanium films: Determination of the chemical composition of the oxide and the oxygen diffusion factor. *Crystals* **2020**, *10*, 117. [CrossRef]
27. Tayade, R.J.; Natarajan, T.S.; Bajaj, H.C. Photocatalytic degradation of methylene blue dye using ultraviolet light emitting diodes. *Ind. Eng. Chem. Res.* **2009**, *48*, 10262–10267. [CrossRef]
28. Hasan, M.M.; Haseeb, A.; Saidur, R.; Masjuki, H.H.; Hamdi, M. Influence of substrate and annealing temperatures on optical properties of RF-sputtered TiO$_2$ thin films. *Opt. Mater.* **2010**, *32*, 690–695. [CrossRef]
29. Chung, Y.-L.; Gan, D.-S.; Ou, K.-L.; Chiou, S.-Y. Formation of anatase and TiO from Ti thin film after anodic treatment and thermal annealing. *J. Electrochem. Soc.* **2011**, *158*, C319. [CrossRef]
30. Sekhar, M.C.; Kondaiah, P.; Jagadeesh Chandra, S.V.; Mohan Rao, G.; Uthanna, S. Effect of substrate bias voltage on the structure, electric and dielectric properties of TiO$_2$ thin films by DC magnetron sputtering. *Appl. Surf. Sci.* **2011**, *258*, 1789–1796. [CrossRef]

31. Sekhar, M.C.; Kondaiah, P.; Radha Krishna, B.; Uthanna, S. Effect of oxygen partial pressure on the electrical and optical Properties of DC magnetron sputtered amorphous TiO_2 films. *J. Spectrosc.* **2013**, *1*, 3–10. [CrossRef]
32. Horprathum, M.; Eiamchai, P.; Chindaudom, P.; Pokaipisit, A.; Limsuwan, P. Oxygen partial pressure dependence of the properties of TiO_2 thin films deposited by DC reactive magnetron sputtering. *Procedia Eng.* **2012**, *32*, 676–682. [CrossRef]
33. Güzelçimen, F.; Tanören, B.; Çetinkaya, Ç.; Kaya, M.D.; Efkere, H.İ.; Özen, Y.; Bingöl, D.; Sirkeci, M.; Kınacı, B.; Ünlü, M.B.; et al. The Effect of thickness on surface structure of Rf sputtered TiO_2 thin films by XPS, SEM/EDS, AFM and SAM. *Vacuum* **2020**, *182*, 109766. [CrossRef]
34. Bharti, B.; Kumar, S.; Lee, H.-N.; Kumar, R. Formation of Oxygen Vacancies and Ti^{3+} State in TiO_2 Thin Film and Enhanced Optical Properties by Air Plasma Treatment. *Sci. Rep.* **2016**, *6*, 32355. [CrossRef]
35. Greczynski, G.; Hultman, L. X-Ray Photoelectron spectroscopy: Towards reliable binding energy referencing. *Prog. Mater. Sci.* **2020**, *107*, 100591. [CrossRef]
36. Fakhouri, H.; Arefi-Khonsari, F.; Jaiswal, A.K.; Pulpytel, J. Enhanced visible light photoactivity and charge separation in TiO_2/TiN bilayer thin films. *Appl. Catal. A Gen.* **2015**, *492*, 83–92. [CrossRef]
37. Gouttebaron, R.; Cornelissen, D.; Snyders, R.; Dauchot, J.P.; Wautelet, M.; Hecq, M. XPS study of TiOx thin films prepared by Dc magnetron sputtering in Ar–O_2 gas mixtures. *Surf. Interface Anal. Int. J. Devoted Dev. Appl. Tech. Anal. Surf. Interfaces Thin Film* **2000**, *30*, 527–530. [CrossRef]
38. Sério, S.; Jorge, M.E.M.; Maneira, M.J.P.; Nunes, Y. Influence of O_2 partial pressure on the growth of nanostructured anatase phase TiO_2 thin films prepared by DC reactive magnetron sputtering. *Mater. Chem. Phys.* **2011**, *126*, 73–81. [CrossRef]
39. Butt, M.A.; Fomchenkov, S.A. Thermal effect on the optical and morphological properties of TiO_2 thin films obtained by Annealing a Ti metal layer. *J. Korean Phys. Soc.* **2017**, *70*, 169–172. [CrossRef]
40. Mechiakh, R.; Meriche, F.; Kremer, R.; Bensaha, R.; Boudine, B.; Boudrioua, A. TiO_2 thin films prepared by Sol–Gel method for waveguiding applications: Correlation between the structural and optical properties. *Opt. Mater.* **2007**, *30*, 645–651. [CrossRef]
41. Peerakiatkhajohn, P.; Butburee, T.; Sul, J.-H.; Thaweesak, S.; Yun, J.-H. Efficient and rapid photocatalytic degradation of methyl orange dye using Al/ZnO nanoparticles. *Nanomaterials* **2021**, *11*, 1059. [CrossRef]
42. Wahab, A.K.; Ould-Chikh, S.; Meyer, K.; Idriss, H. On the "Possible" synergism of the different phases of TiO_2 in photo-catalysis for hydrogen production. *J. Catal.* **2017**, *352*, 657–671. [CrossRef]
43. Wang, P.; Zhou, Q.; Xia, Y.; Zhan, S.; Li, Y. Understanding the charge separation and transfer in mesoporous carbonate-doped phase-junction TiO_2 nanotubes for photocatalytic hydrogen production. *Appl. Catal. B Environ.* **2018**, *225*, 433–444. [CrossRef]
44. Kalarivalappil, V.; Vijayan, B.K.; Kumar, V. Engineering nanocrystalline titania thin films for high photocatalytic activity. *Mater. Today Proc.* **2019**, *9*, 621–626. [CrossRef]
45. Lu, Y.; Matsuzaka, K.; Hao, L.; Hirakawa, Y.; Yoshida, H.; Pan, F.S. Fabrication and photocatalytic activity of TiO_2/Ti composite films by mechanical coating technique and high-temperature oxidation. In Proceedings of the ECCM15—15th European Conference on Composite Materials, Venice, Italy, 24–28 June 2012.
46. Bikondoa, O.; Pang, C.L.; Ithnin, R.; Muryn, C.A.; Onishi, H.; Thornton, G. Direct Visualization of defect-mediated dissociation of water on TiO_2. *Nat. Mater.* **2006**, *5*, 189–192. [CrossRef]
47. Minato, T.; Sainoo, Y.; Kim, Y.; Kato, H.S.; Aika, K.; Kawai, M.; Zhao, J.; Petek, H.; Huang, T.; He, W. The electronic structure of oxygen atom vacancy and hydroxyl impurity defects on titanium dioxide (110) surface. *J. Chem. Phys.* **2009**, *130*, 124502. [CrossRef]
48. Shi, H.; Liu, Y.-C.; Zhao, Z.-J.; Miao, M.; Wu, T.; Wang, Q. Reactivity of the defective rutile TiO_2 surfaces with two bridging-oxygen vacancies: Water molecule as a probe. *J. Phys. Chem. C* **2014**, *118*, 20257–20263. [CrossRef]
49. Elahifard, M.; Sadrian, M.R.; Mirzanejad, A.; Behjatmanesh-Ardakani, R.; Ahmadvand, S. Dispersion of defects in TiO_2 semiconductor: Oxygen vacancies in the bulk and surface of rutile and anatase. *Catalysts* **2020**, *10*, 397. [CrossRef]
50. Mills, A.; Le Hunte, S. An Overview of semiconductor photocatalysis. *J. Photochem. Photobiol. A Chem.* **1997**, *108*, 1–35. [CrossRef]
51. Pan, X.; Yang, M.-Q.; Fu, X.; Zhang, N.; Xu, Y.-J. Defective TiO_2 with oxygen vacancies: Synthesis, properties and photocatalytic applications. *Nanoscale* **2013**, *5*, 3601–3614. [CrossRef] [PubMed]
52. Stojadinović, S.; Tadić, N.; Radić, N.; Grbić, B.; Vasilić, R. Effect of Tb^{3+} doping on the photocatalytic activity of TiO_2 coatings formed by plasma electrolytic oxidation of titanium. *Surf. Coat. Technol.* **2018**, *337*, 279–289. [CrossRef]

Article

Antiferromagnetic Oxide Thin Films for Spintronic Applications

Saima Afroz Siddiqui [1], Deshun Hong [2], John E. Pearson [2] and Axel Hoffmann [1,*]

[1] Materials Research Laboratory, Department of Materials Science and Engineering, University of Illinois at Urbana Champaign, Urbana, IL 61801, USA; saimas@illinois.edu
[2] Materials Science Division, Argonne National Laboratory, Lemont, IL 60439, USA; dhong@anl.gov (D.H.); pearson@anl.gov (J.E.P.)
* Correspondence: axelh@illinois.edu

Abstract: Antiferromagnetic oxides have recently gained much attention because of the possibility to manipulate electrically and optically the Néel vectors in these materials. Their ultrafast spin dynamics, long spin diffusion length and immunity to large magnetic fields make them attractive candidates for spintronic applications. Additionally, there have been many studies on spin wave and magnon transport in single crystals of these oxides. However, the successful applications of the antiferromagnetic oxides will require similar spin transport properties in thin films. In this work, we systematically show the sputtering deposition method for two uniaxial antiferromagnetic oxides, namely Cr_2O_3 and α-Fe_2O_3, on A-plane sapphire substrates, and identify the optimized deposition conditions for epitaxial films with low surface roughness. We also confirm the antiferromagnetic properties of the thin films. The deposition method developed in this article will be important for studying the magnon transport in these epitaxial antiferromagnetic thin films.

Keywords: chromium oxide; hematite; reactive magnetron sputtering; epitaxial thin film; roughness; antiferromagnetic oxides

1. Introduction

Antiferromagnets have gained renewed interest due to their capability to support spin currents via their magnon excitations. This was first recognized in spin pumping experiments, where it was observed that spin currents can be conducted through much thicker insulating antiferromagnetic layers than conventional dielectric materials [1,2]. Subsequently, it was shown that the magnon contribution to heat currents in insulating antiferromagnets can give rise to spin Seebeck effects and therefore can be used to inject spin currents into adjacent metallic layers [3–5]. While this thermal spin current injection relied on incoherent magnons, it has in the meantime also been shown that spin current injection from coherently excited magnons is also possible [6,7]. More importantly, nonlocal transport measurements demonstrated the possibility to electrically inject and detect magnons in both Cr_2O_3 [8] and α-Fe_2O_3 [9,10]. Magnons can propagate over micrometer distances in both of these materials. Lastly, it has also been shown that current induced inhomogeneous temperature profiles can give rise to strains and thus allow to manipulate the magnetic structure within the insulating antiferromagnets [11]. Thus, antiferromagnetic insulators and specifically oxides are promising materials for spintronics applications (i.e., logic, memory, thermoelectric etc.) as they have zero resistive loss, tera-hertz spin dynamics and are immune to high magnetic fields [12,13]. In order to realize their full potential these applications require these antiferromagnetic materials in the thin film form [14].

In this article, we show the systematic variation of the thin films properties (film roughness and strain) of two antiferromagnetic hexagonal materials (i.e., Cr_2O_3 and α-Fe_2O_3 (Hematite)) by varying the deposition parameters (i.e., O_2 flow rate, deposition temperature and deposition pressure). There have been a few studies on the deposition of these oxides in thin film form [15–17], in particular, on (0001) sapphire. We here show the epitaxial growth of these antiferromagnetic thin films on [11$\bar{2}$0] sapphire substrates. These antiferromagnetic materials show no residual magnetism within the thickness ranging from 10 to 200 nm. However, the Morin transition was observed in α-Fe_2O_3 with thickness above 200 nm. The deposition process developed in this article will enable many exciting spintronic studies in these antiferromagnetic oxides.

2. Deposition Methods and Characterizations

In this work, we deposited Cr_2O_3 and α-Fe_2O_3 by using radio frequency (RF) magnetron sputtering at high temperature from 2″ Cr and Fe sputtering targets, respectively, on (11–20) Al_2O_3 (A-plane sapphire) substrates in the presence of both oxygen and argon. Initially, we varied the oxygen flow rate from 1.0 to 5.0 sccm while keeping the argon flow rate at 70 sccm and the chamber pressure at 3 mTorr. Here, we calibrated the deposition rate of Cr_2O_3 and the α-Fe_2O_3 using the crystal monitor. The temperature of the substrate was varied from 625 to 750 °C and the deposition pressure was varied from 2 to 5 mTorr for both Cr_2O_3 and α-Fe_2O_3.

The crystal orientation and the deposition rate of the epitaxial oxide films are characterized by X-ray diffraction (XRD) and X-ray reflectivity (XRR), respectively, using the Bruker D8 Advance XRD System (Bruker Corporation, Billerica, MA, USA) with a monochromatic Cu Kα source with a wavelength of 1.54 Å. The roughness of the films is characterized by both XRR and atomic force microscopy (AFM) and the magnetic properties are characterized by superconducting quantum interference device (SQUID) magnetometer (Quantum Design North America, San Diego, CA, USA).

3. Results and Discussions

3.1. Crystal Structures of Epitaxial Oxide Films

Figure 1 shows the XRD data of the 20 nm Cr_2O_3 epitaxial films grown at 675 °C with an oxygen flow of 2.5 sccm and chamber pressure of 5 mTorr. The base pressure of the chamber was 3×10^{-8} Torr. After the deposition, the films were annealed at 700 °C for one hour at 2 mTorr in the presence of oxygen at 1 mTorr. XRD θ–2θ patterns of film in Figure 1a show the peak at 2θ = 36.02°, which comes from the Cr_2O_3 [11–20] Bragg reflection corresponding to the corundum structure. The peak at 37.8° is the [11–20] Al_2O_3 substrate peak. The films grown at oxygen flow varying from 2.0 to 3.5 sccm show similar structure with different strains in the films as discussed in detail below. It was, however, found that when the amount of O_2 in the gas mixture was above 5.0 sccm, the deposition rate decreased sharply. This finding is consistent with the previously reported results [18].

Figure 1b shows the rocking curve data for the [11–20] Bragg reflection of the same 20-nm Cr_2O_3 sample. The full width at half maximum value of the rocking curve is 0.145°. This indicates that the Cr_2O_3 film is formed with conformal a-axis orientation on the Al_2O_3 substrate. To confirm the epitaxial state of the in-plane orientation of the film, we performed the XRD Φ-scan of the Cr_2O_3 film from −190° to 190° (see Figure 1c). The Φ-scanning result from the Cr_2O_3 films indicates that there are two equivalent peaks, each being separated by 180°. These commensurate peak positions of the film and the substrate confirm the epitaxial relationship between the two. And this two-fold symmetry of the deposited film indicates that the film consists of a single crystalline domain.

We deposited both Cr_2O_3 and Fe_2O_3 samples by varying the flow rate of oxygen. The XRD θ–2θ scan of the Cr_2O_3 thin films deposited with different oxygen flow are shown in Figure 2. The thickness of the films deposited at 3.5 sccm is 10 nm, while it is 20 nm for the other films deposited films. For both 2.5 and 3.0 sccm, the Cr_2O_3 peaks show finite size oscillations up to the fifth order, which confirms low surface roughness of these films. However, the film deposited with 3.5 sccm pressure shows a much broader XRD peak. The broadening may result from the shorter out of plane coherence length in thinner film [19]. Thus, for the subsequent depositions, we choose the oxygen flow of 2.5 sccm for all the deposition conditions as this flow of oxygen provides the optimum oxidation for both oxides.

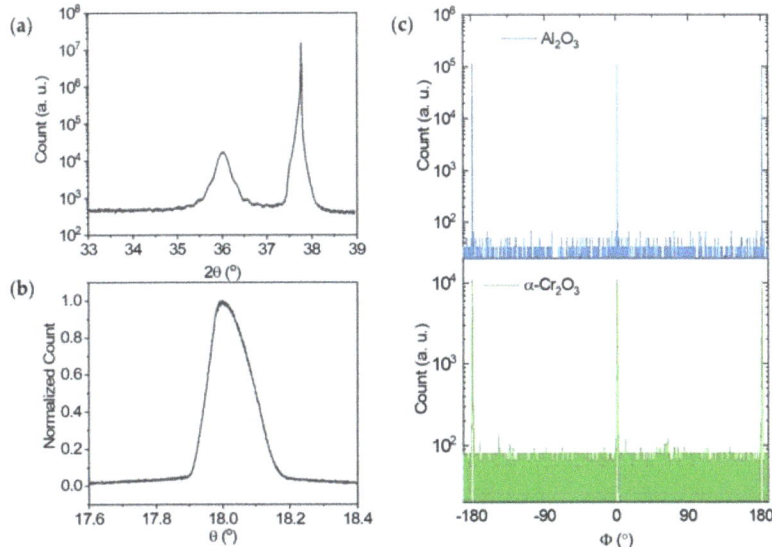

Figure 1. (a) XRD θ–2θ patterns of the Cr_2O_3 epitaxial film grown on an [11$\bar{2}$0] Al_2O_3 substrate. (b) XRD θ-rocking curve Cr_2O_3 film. (c) XRD Φ-scan results of the Cr_2O_3 and Al_2O_3 reflections.

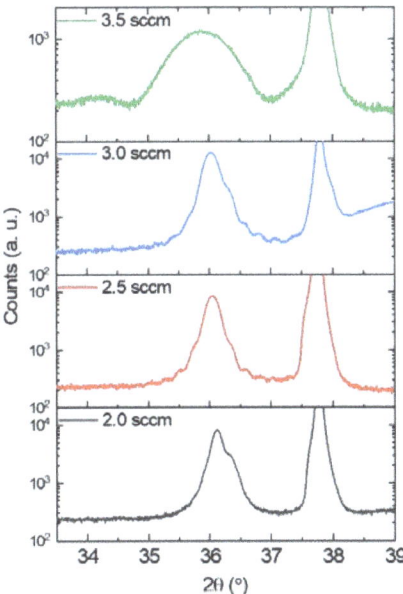

Figure 2. θ–2θ scan of Cr_2O_3 thin films deposited with different oxygen flow at a deposition pressure of 5 mTorr and temperature of 675 °C with 150 W RF power.

We characterize the roughness of the Cr_2O_3 epitaxial films using XRR. Figure 3a,b shows the roughness at different deposition temperatures as a function of O_2 flow and deposition pressure, respectively. All the Cr_2O_3 films shown in Figure 3a were deposited with a chamber pressure of 5 mTorr. The roughness of the Cr_2O_3 films is lowest at the oxygen flow rate of 2.5 sccm and the deposition temperature of 675 °C. The roughness of the films increases for both lower and higher oxygen flow rate. Figure 3a also shows the in-plane compressive strain in Cr_2O_3 deposited at 675 °C due to lattice mismatch between the thin films and the sapphire substrate. Figure 3b shows that the roughness of the films gradually decreases for higher deposition pressure. We limit the deposition pressure to 5 mTorr to obtain an epitaxial films film with low surface roughness while the deposition rate is still reasonable (3.5 Å/s). The low surface roughness of these antiferromagnet oxide films is important for the low damping of the magnon modes [20].

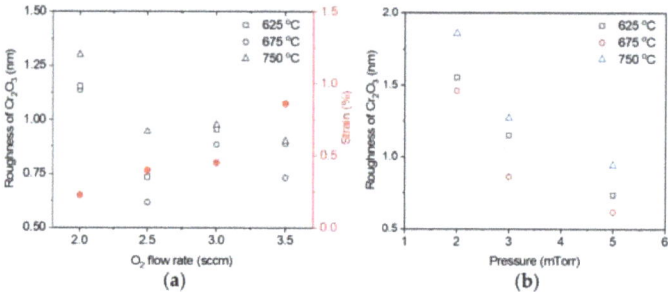

Figure 3. (a) Surface roughness of Cr_2O_3 films with thickness of 20 nm deposited at different temperature and in-plane compressive strain in Cr_2O_3 at 675 °C as a function of oxygen flow rate. The open symbols show the surface roughness, and the filled circles show the in-plane strain in Cr_2O_3 films. (b) Surface roughness of the films as a function of deposition pressure.

Figure 4 shows the XRD data of a 200-nm Fe_2O_3 film on an A-plane sapphire substrate. The peak at 35.6° is the [11–20] peak of α-Fe_2O_3. This film is capped with 5 nm of Pt. The broader peak at 39.5° shows the [111] XRD peak of the Pt layer. It was deposited on Fe_2O_3 at room temperature at a deposition pres × 3 μm area and found to be only 3 Å (Figure 5). The in-plane Φ XRD-scan of the α-Fe_2O_3 film indicates two-fold symmetry along the <11$\bar{2}$0> direction, and therefore, single crystalline domain structure. The roughnesses of thinner α-Fe_2O_3 films show a similar trend with the oxygen flow and deposition pressure as observed for the Cr_2O_3.

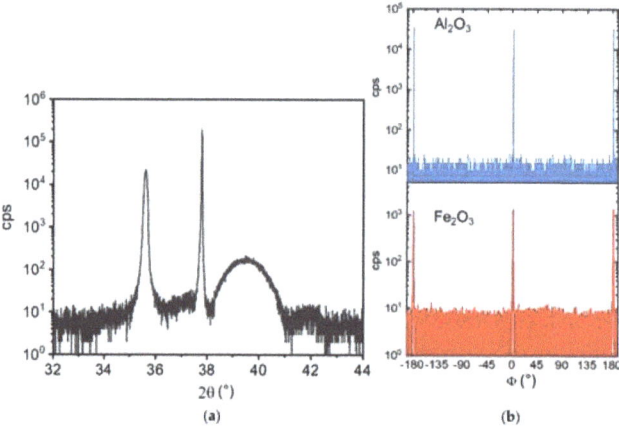

Figure 4. (a) XRD θ–2θ patterns of the α-Fe_2O_3 epitaxial film grown on an [11$\bar{2}$0] Al_2O_3 substrate. (b) XRD Φ-scan results of the α-Fe_2O_3 and Al_2O_3 reflections.

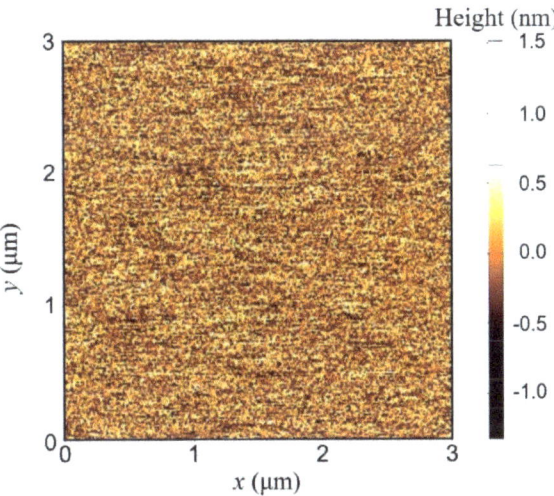

Figure 5. Atomic force micrograph of 200 nm thick α-Fe$_2$O$_3$ film.

3.2. Magnetic Properties

Cr$_2$O$_3$ and α-Fe$_2$O$_3$ are both antiferromagnetic materials. To confirm the absence of any residual magnetism in Cr$_2$O$_3$ and α-Fe$_2$O$_3$ films, SQUID measurements were performed with magnetic fields of 0.4 and 0.2 T, respectively, applied perpendicular to the A-plane of the substrates. For the SQUID measurement, a 5 × 5 mm sample is loaded in the chamber inside a straw. The sample plane is perpendicular to the magnetic field. Figure 6 shows the magnetization of the epitaxial oxide films as a function of temperature. There is no hysteresis present in the magnetization data between the heating and cooling of the Cr$_2$O$_3$ films from 5 to 320 K and, of the α-Fe$_2$O$_3$ films from 5 to 350 K. For the 20-nm Cr$_2$O$_3$ films, the magnetization is almost zero, which identifies the antiferromagnetic exchange interaction between Cr^{3+} ions in neighboring layers [21]. There is no significant difference observed between the zero-field cooled (ZFC) and the field cooled (FC) measurements. For the FC measurement, the sample is cooled to 5 K from 320 K with a magnetic field of 2 T applied perpendicular to the A-plane of the substrate. We noticed a very small (0.01 emu/cm^3) increase in the magnetization at 40 K for 20-nm Cr$_2$O$_3$ epitaxial films. This minuscule magnetization comes from the magnetic impurities present in the sapphire substrates as confirmed by measuring the magnetization of only the substrate as a function of temperature (data not presented). The inset of Figure 6a shows the spin orientation of the Cr^{3+} ions inside a unit cell of Cr$_2$O$_3$. The spins are pointing along the c-axis of the sample for all temperatures.

In α-Fe$_2$O$_3$, the antiferromagnetic spin configuration changes its direction from being parallel to the [0001] axis to being in the (0001) basal plane at temperatures above the Morin temperature (T_M). The Morin transition is due to the Dzyaloshinskii–Moriya interaction, where asymmetric exchange interaction between two neighboring spins results in a weak net magnetic moment in the (0001) plane at temperatures above T_M [22]. Figure 4b shows the temperature dependence of the out-of-plane magnetization of a 200-nm thick α-Fe$_2$O$_3$ film under an applied field of 0.2 T. An increase of the magnetization is observed above 225 K, which is lower than in bulk samples (≈260 K) [22]. The enhanced magnetization reflects a weak net ferromagnetic moment at temperatures above T_M. For thinner α-Fe$_2$O$_3$ samples, we did not observe the Morin transition. The lowering of the transition temperature in thinner films may result from the increased in-plane strain in those films. The Morin transition is usually determined by the competition between the magnetic dipolar and the structural anisotropy energies. The temperature variations of these two energy contributions have quantum statistical basis and are different [23]. It is possible that their temperature variations change compared to bulk crystals because of the presence of strain in the film, which eventually eliminates the secondary transition.

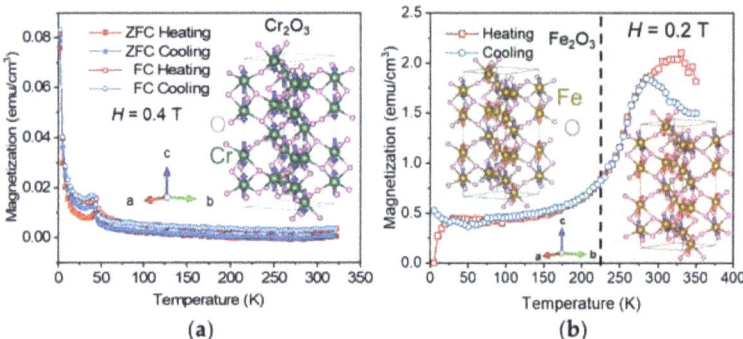

Figure 6. Magnetization of (**a**) Cr_2O_3, (**b**) α-Fe_2O_3 epitaxial films along the a-axis as a function of temperature. A magnetic field of 2 T is applied perpendicular to the A-plane during the field-cooling for the Cr_2O_3 film. The inset of (**a**) shows the unit cell of Cr_2O_3 with the spin orientations for the Cr^{3+} ions. The insets of (**b**) show the unit cells of α-Fe_2O_3 with the spin orientations for the Fe^{3+} ions before and after the Morin transition.

4. Conclusions

In summary, we have shown a detailed study of radio frequency magnetron sputtering of epitaxial thin films of Cr_2O_3 and α-Fe_2O_3 on A-plane sapphire substrates for spintronic applications. Optimized sputtering conditions are discussed together with the characterizations of the crystal orientations and magnetic properties of these films. These single domain antiferromagnetic oxide thin films will be very important for studying the magnon dynamics and transport for extremely low-loss spintronic devices. They can also potentially be used for antiferromagnetic ultra-dense memory with THz bandwidth by utilizing current induced magnetization reversal. In addition, our developed deposition method confirms the single domain properties of these films, which is essential to reproduce many exciting magnon properties, which were previously observed only in single crystal oxides. Our epitaxial sputtering method will pave the way for the potential application of antiferromagnetic oxides in the field of electronics.

Author Contributions: S.A.S. conceived, designed and performed the experiments, and analyzed the data. D.H. helped with the α-Fe_2O_3 structural characterization. J.E.P. helped with the deposition and characterization of the α-Fe_2O_3 films. S.A.S. and A.H. co-wrote the paper. All authors have read and agreed to the published version of the manuscript.

Funding: The growth and the characterization of α-Fe_2O_3 films were supported by the US Department of Energy, Office of Science, Basic Energy Sciences, Materials Sciences and Engineering Division. The use of facilities at the Center for Nanoscale Materials, an Office of Science user facility, was supported by the US Department of Energy, Basic Energy Sciences under Contract No. DE-AC02-06CH11357. The growth and the characterization of Cr_2O_3 films were supported by the National Science Foundation (NSF) through the University of Illinois at Urbana-Champaign Materials Research Science and Engineering Center No. DMR-1720633 and was carried out in part in the Materials Research Laboratory Central Research Facilities, University of Illinois.

Institutional Review Board Statement: Not applicable.

Informed Consent Statement: Not applicable.

Data Availability Statement: The data presented in this study are available on request from the corresponding author.

Conflicts of Interest: The authors declare no conflict of interest. The funders had no role in the design of the study; in the collection, analyses, or interpretation of data; in the writing of the manuscript, or in the decision to publish the results.

References

1. Wang, H.; Du, C.; Hammel, P.C.; Yang, F. Antiferromagnonic spin transport from $Y_3Fe_5O_{12}$ into NiO. *Phys. Rev. Lett.* **2014**, *113*, 097202. [CrossRef]
2. Hahn, C.; de Loubens, G.; Naletov, V.V.; Ben Youssef, J.; Klein, O.; Viret, M. Conduction of spin currents through insulating antiferromagnetic oxides. *EPL Europhys. Lett.* **2014**, *108*, 57005. [CrossRef]
3. Seki, S.; Ideue, T.; Kubota, M.; Kozuka, Y.; Takagi, R.; Nakamura, M.; Kaneko, Y.; Kawasaki, M.; Tokura, Y. Thermal generation of spin current in an antiferromagnet. *Phys. Rev. Lett.* **2015**, *115*, 266601. [CrossRef]
4. Wu, S.M.; Zhang, W.; Kc, A.; Borisov, P.; Pearson, J.E.; Jiang, J.S.; Lederman, D.; Hoffmann, A.; Bhattacharya, A. Antiferromagnetic spin seebeck effect. *Phys. Rev. Lett.* **2016**, *116*, 097204. [CrossRef]
5. Luo, Y.; Liu, C.; Saglam, H.; Li, Y.; Zhang, W.; Zhang, S.S.; Pearson, J.E.; Fisher, B.; Zhou, T.; Bhattacharya, A.; et al. Distinguishing antiferromagnetic spin sublattices via the spin Seebeck effect. *Phys. Rev. B* **2021**, *103*, L020401. [CrossRef]
6. Li, J.; Wilson, C.B.; Cheng, R.; Lohmann, M.; Kavand, M.; Yuan, W.; Aldosary, M.; Agladze, N.; Wei, P.; Sherwin, M.S.; et al. Spin current from sub-terahertz-generated antiferromagnetic magnons. *Nature* **2020**, *578*, 70–74. [CrossRef]
7. Vaidya, P.; Morley, S.A.; van Tol, J.; Liu, Y.; Cheng, R.; Brataas, A.; Lederman, D.; del Barco, E. Subterahertz spin pumping from an insulating antiferromagnet. *Science* **2020**, *368*, 160. [CrossRef]
8. Yuan, W.; Zhu, Q.; Su, T.; Yao, Y.; Xing, W.; Chen, Y.; Ma, Y.; Lin, X.; Shi, J.; Shindou, R.; et al. Experimental signatures of spin superfluid ground state in canted antiferromagnet Cr_2O_3 via nonlocal spin transport. *Sci. Adv.* **2018**, *4*, eaat1098. [CrossRef]
9. Lebrun, R.; Ross, A.; Bender, S.A.; Qaiumzadeh, A.; Baldrati, L.; Cramer, J.; Brataas, A.; Duine, R.A.; Kläui, M. Tunable long-distance spin transport in a crystalline antiferromagnetic iron oxide. *Nature* **2018**, *561*, 222–225. [CrossRef]
10. Lebrun, R.; Ross, A.; Gomonay, O.; Baltz, V.; Ebels, U.; Barra, A.L.; Qaiumzadeh, A.; Brataas, A.; Sinova, J.; Klaui, M. Long-distance spin-transport across the Morin phase transition up to room temperature in ultra-low damping single crystals of the antiferromagnet α-Fe_2O_3. *Nat. Commun.* **2020**, *11*, 6332. [CrossRef]
11. Zhang, P.; Finley, J.; Safi, T.; Liu, L. Quantitative study on current-induced effect in an antiferromagnet insulator/Pt bilayer film. *Phys. Rev. Lett.* **2019**, *123*, 247206. [CrossRef]
12. Kosub, T.; Kopte, M.; Hühne, R.; Appel, P.; Shields, B.; Maletinsky, P.; Hübner, R.; Liedke, M.O.; Fassbender, J.; Schmidt, O.G.; et al. Purely antiferromagnetic magnetoelectric random access memory. *Nat. Commun.* **2017**, *8*, 13985. [CrossRef]
13. Ramazanov, S.; Sobola, D.; Orudzhev, F.; Knápek, A.; Polčák, J.; Potoček, M.; Kaspar, P.; Dallaev, R. Surface modification and enhancement of ferromagnetism in $BiFeO_3$ nanofilms deposited on HOPG. *Nanomaterials* **2020**, *10*, 1990. [CrossRef] [PubMed]
14. Speriosu, V.S.; Herman, D.A.; Sanders, I.L.; Yogi, T. Magnetic thin films in recording technology. *IBM J. Res. Dev.* **1990**, *34*, 884–902. [CrossRef]
15. Leighton, C.; Hoffmann, A.; Fitzsimmons, M.R.; Nogués, J.; Schuller, I.K. Deposition of epitaxial α-Fe_2O_3 layers for exchange bias studies by reactive dc magnetron sputtering. *Philos. Mag. B* **2001**, *81*, 1927–1934. [CrossRef]
16. Valeri, S.; Altieri, S.; Luches, P. Growth of antiferromagnetic oxide thin films. *Magn. Prop. Antiferromagn. Oxide Mater.* **2010**, 25–68. [CrossRef]
17. Shimomura, N.; Pati, S.P.; Sato, Y.; Nozaki, T.; Shibata, T.; Mibu, K.; Sahashi, M. Morin transition temperature in (0001)-oriented α-Fe_2O_3 thin film and effect of Ir doping. *J. Appl. Phys.* **2015**, *117*, 17C736. [CrossRef]
18. Rothhaar, U.; Oechsner, H. Rf magnetron sputter deposition of Cr_2O_3 layers on ceramic Al_2O_3 substrates. *Surf. Coat. Technol.* **1993**, *59*, 183–186. [CrossRef]
19. Vayunandana Reddy, Y.K.; Wolfman, J.; Autret-Lambert, C.; Gervais, M.; Gervais, F. Strain relaxation of epitaxial $(Ba_{0.6}Sr_{0.4})(Zr_{0.3}Ti_{0.7})O_3$ thin films grown on $SrTiO_3$ substrates by pulsed laser deposition. *J. Appl. Phys.* **2010**, *107*, 106101.
20. Yu, T.; Sharma, S.; Blanter, Y.M.; Bauer, G.E. Surface dynamics of rough magnetic films. *Phys. Rev. B* **2019**, *99*, 174402. [CrossRef]
21. Nozaki, T.; Al-Mahdawi, M.; Shiokawa, Y.; Pati, S.P.; Imamura, H.; Sahashi, M. Magnetic anisotropy of doped Cr_2O_3 antiferromagnetic films evaluated by utilizing parasitic magnetization. *J. Appl. Phys.* **2020**, *128*, 023901. [CrossRef]
22. Morin, F.J. Electrical properties of αFe_2O_3 and αFe_2O_3 containing titanium. *Phys. Rev.* **1951**, *83*, 1005–1010. [CrossRef]
23. Artman, J.O.; Murphy, J.C.; Foner, S. Magnetic anisotropy in antiferromagnetic corundum-type sesquioxides. *Phys. Rev.* **1965**, *138*, A912–A917. [CrossRef]

Article

Structural, Magnetic and Gas Sensing Activity of Pure and Cr Doped In$_2$O$_3$ Thin Films Grown by Pulsed Laser Deposition

Veeraswamy Yaragani [1,*,†], Hari Prasad Kamatam [1,2,†], Karuppiah Deva Arun Kumar [3], Paolo Mele [3,*], Arulanandam Jegatha Christy [4], Kugalur Venkidusamy Gunavathy [5], Sultan Alomairy [6] and Mohammed Sultan Al-Buriahi [7]

[1] Department of Physics, Institute of Aeronautical Engineering, Dundigal, Hyderabad 500043, India; k.hariprasad500@gmail.com
[2] Department of Physics, Pondicherry University, Puducherry 605014, India
[3] College of Engineering, Shibaura Institute of Technology, Saitama 337-8570, Japan; i048267@shibaura-it.ac.jp
[4] Department of Physics, Jayaraj Annapackiam College for Women, Periyakulam, Theni 625601, India; jegathachristy@gmail.com
[5] Department of Physics, Kongu Engineering College, Perundurai 638060, India; gunavathy@kongu.ac.in
[6] Department of Physics, College of Science, Taif University, P.O. Box 11099, Taif 21944, Saudi Arabia; mohsaa996@gmail.com
[7] Department of Physics, Sakarya University, 54100 Sakarya, Turkey; mohammed.al-buriahi@ogr.sakarya.edu.tr
* Correspondence: veerayaragani@gmail.com (V.Y.); pmele@shibaura-it.ac.jp (P.M.)
† Author contributed equally.

Abstract: Pure In$_2$O$_3$ and 6% Cr-doped In$_2$O$_3$ thin films were prepared on a silicon (Si) substrate by pulsed laser deposition technique. The obtained In$_2$O$_3$/In$_2$O$_3$:Cr thin films structural, morphological, optical, magnetic and gas sensing properties were briefly investigated. The X-ray diffraction results confirmed that the grown thin films are in single-phase cubic bixbyte structure with space group Ia-3. The SEM analysis showed the formation of agglomerated spherical shape morphology with the decreased average grain size for Cr doped In$_2$O$_3$ thin film compared to pure In$_2$O$_3$ film. It is observed that the Cr doped In$_2$O$_3$ thin film shows the lower band gap energy and that the corresponding transmittance is around 80%. The X-ray photoelectron spectroscopy measurements revealed that the presence of oxygen vacancy in the doped In$_2$O$_3$ film. These oxygen defects could play a significant role to enhance the sensing performance towards chemical species. In the magnetic hysteresis loop, it is clear that the prepared films confirm the ferromagnetic behaviour and the maximum saturation value of 39 emu/cc for Cr doped In$_2$O$_3$ film. NH$_3$ gas sensing studies was also carried out at room temperature for both pure and Cr doped In$_2$O$_3$ films, and the obtained higher sensitivity is 182% for Cr doped In$_2$O$_3$, which is about nine times higher than for the pure In$_2$O$_3$ film due to the presence of defects on the doped film surface.

Keywords: In$_2$O$_3$/In$_2$O$_3$:Cr thin films; XPS; magnetization property; NH$_3$ sensor

Citation: Yaragani, V.; Kamatam, H.P.; Deva Arun Kumar, K.; Mele, P.; Christy, A.J.; Gunavathy, K.V.; Alomairy, S.; Al-Buriahi, M.S. Structural, Magnetic and Gas Sensing Activity of Pure and Cr Doped In$_2$O$_3$ Thin Films Grown by Pulsed Laser Deposition. *Coatings* **2021**, *11*, 588. https://doi.org/10.3390/coatings11050588

Academic Editor: Joe Sakai

Received: 18 April 2021
Accepted: 14 May 2021
Published: 17 May 2021

Publisher's Note: MDPI stays neutral with regard to jurisdictional claims in published maps and institutional affiliations.

Copyright: © 2021 by the authors. Licensee MDPI, Basel, Switzerland. This article is an open access article distributed under the terms and conditions of the Creative Commons Attribution (CC BY) license (https://creativecommons.org/licenses/by/4.0/).

1. Introduction

High carrier mobility with the magnetic property of transparent metal oxides attracts as a miniature robust device for spintronic applications. This class of materials is called diluted magnetic metal oxide semiconductors (DMOS) [1]. These magnetic semiconductors could lead to unite the electrical manipulation of magnetic states and the magnetic adjustment of electrical signals that could result in devices such as bipolar transistors, spin resonant diodes, spin field effect transistors, magnetic semiconductor tunnel junction devices, magnetic bipolar junction diodes, and transistors, etc. [2–8]. If carrier-mediated magnetization can be induced in transparent semiconducting oxides such as In$_2$O$_3$, ZnO, TiO$_2$, etc., it is predicted that such DMOS will exhibit coupling among electrical, optical, and magnetic properties, further boosting the prospects of devices emanating from such materials. Among various transparent conducting oxide materials, In$_2$O$_3$, ZnO and SnO$_2$

are multipurpose materials with a wide range of applications in optoelectronics, solar cells, gas sensors, etc. [9–13], due to the interstitial defects of oxygen and other state defects. In_2O_3 is an n-type semiconductor with high optical bandgap with cubic bixbite crystal structure. Inducing magnetic ordering in In_2O_3 based materials will further enhance its utility in several devices. In the past, a few attempts were made to probe the magnetic properties of transition metal doped In_2O_3. However, the origin of magnetism in such system has been a matter of debate. In those, Philip et al. [14] reported that ferromagnetic ordering in Cr doped In_2O_3 is due to carrier mediation. Jayakumar et al. [15] suggested that defects in the synthesis may cause magnetism in Fe-doped In_2O_3 thin films. Garcia et al. [16] reported that double exchange interaction between metal ions of different valence oxides. According to Chang et al., Mo doped In_2O_3 exhibited good ferromagnetism attributed to the indirect exchange interaction of the charge carriers available in In_2O_3 [17]. Khare et al. discussed the origin of room temperature ferromagnetism in Cr doped In_2O_3 films and they observed that the ferromagnetic behaviour can improve after high vacuum annealing [18]. Ukah et al. described the structural and electrical transport properties of Cr doped In_2O_3 films and they suggested that the pulsed laser deposition technique (PLD) has more advantage than other deposition techniques [19]. From the above discussions, it is evident that the properties of In_2O_3 based DMOS depend on the dopant; and, defects would alter the electronic structure of such materials. However, not much is known about the modification in the electronic structure of In_2O_3 occurring due to such defects or dopants. Probing the electronic structure will be key to understanding the underlying mechanism responsible for the coupled electrical, optical and magnetic properties of In_2O_3 based DMOS and to accordingly design the material for device application. It is also predicted that the transparent semiconducting indium oxide (In_2O_3) based DMOS would exhibit coupling among electrical, optical, and magnetic properties.

In addition, we also investigate the ammonia (NH_3) gas sensing properties of In_2O_3 because of its surface controlled type semiconductor [20], which reacts well with the surrounding chemical species. In the case of metal doped In_2O_3, researchers have mainly focused on magnetic device applications due to the native ferromagnetic properties. In_2O_3 based thin film sensor could be applicable for gas sensing devices due to impurity defects and/or native defects of oxygen vacancies by metal doping and or changing the deposition parameters. Recently, Wang et al. [21] synthesized Co-doped In_2O_3 nanorods for formaldehyde (HCHO) gas sensor. They observed that the increase of response from undoped to doped films is attributed to the doping elements, creating a high surface area which could offer the adsorption of gas molecules and improve the response. Han et al. [22] reported the response of Ce doped In_2O_3 nanospheres towards methanol gas at an operating temperature of 320 °C. They observed the maximum response of ~35 at 100 ppm of methanol for the doped film, which is higher than the pure In_2O_3. Further, they obtained fast response/recovery times (14/10 s) with respect to methanol due to the sensor operating temperature. Manivasaham et al. [23] prepared undoped and Cr doped ZnO films for room temperature NH_3 sensor and obtained the best response for Cr doped ZnO compared to undoped ZnO film. From the above reports, the defects of oxygen play a major role in improving the sensing response towards any chemical species. Among other gases, sensing behaviour is high for ammonia at room temperature. The room temperature sensors have numerous profits such as less complex circuitry, safety in burnable situation, low power consumption, and there is no requirement of any heating circuits, etc.

In this study, we report the structural, magnetic, optical and gas sensing studies of pure In_2O_3 and Cr doped In_2O_3 films fabricated by pulsed laser deposition (PLD). The fabricated In_2O_3/In_2O_3:Cr thin films were characterized by different analytical techniques such as XRD, SEM, EDX, XPS, VSM, UV-visible spectrophotometer, and room temperature NH_3 gas sensor measurements.

2. Materials and Methods

2.1. Pellet Preparation by SSR

A well-sintered pellet of pure In_2O_3 and Cr doped In_2O_3 ($In_{2-x}Cr_xO_3$ (x = 0.06)) were prepared by solid state reaction (SSR) technique. The high purity In_2O_3 and Cr_2O_3 powders (acquired from Sigma Aldrich) were taken as starting materials to prepare the $In_{2-x}Cr_xO_3$ (x = 0.06) powder sample by solid state reaction. Firstly, the required quantity of both In and Cr oxide powders were mixed together for 5 h in a ball milling setup. Further, the mixed powder was continuously heated in air at 800 °C for 10 h and the prepared high purity In_2O_3 and Cr doped In_2O_3 powders were cold pressed at a 10-ton load with the desired dimensions. Finally, the prepared pellets of 25 mm diameter and 2 mm thickness were sintered at 900 °C for 12 h. The density of the sintered pellets was calculated by an immersion-specific gravity method and the value of relative density in percentage was found to be ~96%.

2.2. Film Preparation by PLD

In the present investigation, the prepared high density In_2O_3 and Cr doped In_2O_3 pellets were used to fabricate thin films on Si (100) substrate by pulsed laser deposition (PLD) technique (Excimer laser KrF (λ = 248 nm) source; Lambda Physik COMPex 201 Model, Göttingen, Germany). Prior to the deposition, the native oxide must be removed from the Si substrate. Therefore, we used $HF:H_2O$ (1:10) solution treatment for 2 min and then washed with deionized water thrice. Then, pre-cleaned substrate was dried and cleaned with high pure (99.999%) nitrogen gas before loading into the PLD chamber. Then, the PLD chamber was evacuated to a base vacuum of 10^{-6} Torr. During the deposition, the energy density and pulse repetition rates were adjusted at 2.5 J/cm^2 and 10 Hz, respectively, for all thin films. The detailed experimental conditions are listed in Table 1. Oxygen partial pressure was constantly maintained at 1 mTorr for the entire film deposition. Finally, the prepared films were stored in a desiccator to avoid atmosphere contaminations.

Table 1. PLD experimental conditions for Cr doped In_2O_3 thin films.

Parameters	Conditions
Source	KrF Excimer Laser
Base vacuum	1×10^{-6} Torr
Target to substrate distance	6 cm
Substrate rotation	60 rpm
Substrate temperature	670 °C
Laser wavelength	248 nm
Energy density	2.5 J/cm^2
Pulse repetition rate	10 Hz
Pulse duration	8 ns
Deposition time	15 min

2.3. Characterizations

The structural property of films was carried out using GIXRD (D8-Discover system of M/s Bruker, Billerica, MA, USA) equipped with CuK$_\alpha$. The X-ray photoelectron spectroscopy (XPS) measurement was performed by Omicron energy analyser (EA-125, Vancouver, BC, Canada) to analyse the chemical compositions. All collected data were normalized. The vibrating sample magnetometer (VSM) (Lake Shore: Model: 7404, Westerville, OH, USA) was performed to investigate the magnetic properties of films. The film morphology was analysed using scanning electron microscope (model ZEISS, Oberkochen, Germany). The optical transmittance of thin films was recorded in the wavelength range between 300 and 1000 nm using a Shimadzu Solid Spec-3700 DUV UV-visible spectrophotometer (Kyoto, Japan). The ammonia (NH_3) sensing properties of In_2O_3/In_2O_3:Cr thin films were investigated by sensor setup with the help of Keithley Source Meter (model 2450, Tectronix Inc., Beaverton, OR, USA). All the characterizations were carried out at room temperature.

3. Results and Discussion

3.1. Structural Analysis

Figure 1 shows the XRD patterns of the prepared pure In_2O_3 and Cr doped In_2O_3 thin films. The observed X-ray diffraction peaks match with the JCPDS (card No. 06-0416) standard data file of indium oxide (In_2O_3), confirming the formation of cubic bixbyte crystal structure of In_2O_3 with the Ia-3 space group. No extra characteristic peaks were identified in the films that can be related to any metal chromium, and chromium-based oxide compounds. Pure In_2O_3 and Cr doped In_2O_3 thin films exhibit polycrystalline structure with a strong preferred orientation along (222) plane direction. In addition, the (222) plane peak position was slightly shifted towards the higher angle for Cr doped In_2O_3 thin film attributed to interstitial doping of metallic cations. The lattice parameters are evaluated by analysing the observed XRD patterns of the prepared films using the celref3 software and are found to be 10.11 and 10.12 Å, respectively. The average crystallite size value of pure and Cr doped In_2O_3 thin films were determined using Scherer's equation [24].

$$D = \frac{0.9\,\lambda}{\beta \cos \theta},\qquad(1)$$

where, D is the average crystallite size, β is the full width at half maximum (FWHM) of the diffraction peak, λ is the wavelength of CuK_α radiation, and θ is the Bragg's angle. The calculated average crystallite size values of the pure In_2O_3 and Cr doped In_2O_3 films are found to be 26 and 21 nm, respectively. Clearly, the calculated D value is decreased from undoped to doped In_2O_3 films due to increase of FWHM. Krishna et al. [25] reported the reduction of crystallite size from 40 to 31 nm for pure In_2O_3 and Cr-doped In_2O_3 thin films. The decrease of the D value might be due to the reason that doped Cr ions might disturb the grain growth level of native In-O lattice. From the XRD analysis, it is confirmed that the formation of nanocrystalline cubic bixbyte phase for both pure In_2O_3 and Cr-doped In_2O_3 thin films.

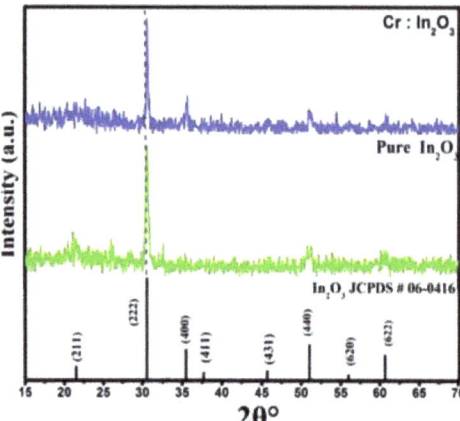

Figure 1. XRD patterns of pure In_2O_3 and Cr doped In_2O_3 thin films with a standard JCPDS file data.

3.2. Surface Morphology Analyses

Figure S1 (Supplementary Materials) shows the SEM micrographs of the fabricated (a) pure In_2O_3 and (c) Cr-doped In_2O_3 thin films, respectively. From the figure, the observed SEM micrographs show the agglomerated spherical shaped grains for both pure and Cr doped In_2O_3 thin films. The average particle (grain) size of pure In_2O_3 and Cr-doped In_2O_3 thin films, respectively, are found to be in the ranges 25–40 nm and 20–35 nm. It is revealed that Cr doped In_2O_3 thin film has less particle size when compared to pure In_2O_3; which is correlated with XRD results. A lesser particle size could enhance the surface to volume ratio; this offers an appropriate increased surface environment for effective adsorption of surrounding gases. Consequently, there is an increase of the gas sensing response with quick reaction time. Figure S1b,d shows the EDX spectrum of pure In_2O_3 and Cr doped In_2O_3 films, respectively. From Figure S1b, EDX spectrum shows the presence of In and O elements in the pure In_2O_3 thin film. The obtained EDX spectrum confirms the absence of other impurities, which established the purity of the In_2O_3. In Figure S1d, EDX spectrum shows the existence of In, Cr and O elements in the Cr doped In_2O_3 thin film and the absence of other elements. Both pure and doped In_2O_3 films exhibit a high intense peak at 1.8 keV, corresponding to the Si element, which is originated from the substrate.

Figure 2 shows the 2D and 3D AFM images of pure In_2O_3 and Cr doped In_2O_3 thin films. The surface roughness is one of the crucial parameters to understand the sensing performance of metal oxide thin films. As seen from Figure 2a, the pure In_2O_3 film shows spherical grains uniformly arranged on the film surface, whereas, as shown in Figure 2b, the Cr doped film exhibits spherical grains. The average surface roughness values were found to be 9 and 14 nm for pure and Cr doped In_2O_3 films, respectively. In general, the rough film surface performs well in sensing behaviour by attracting more oxygen species from the air and releasing the electron with the interaction between oxygen and the surrounding gases. Among the prepared films, Cr doped In_2O_3 film have a higher surface roughness than that for pure In_2O_3, which might be due to the formation of Cr defects in the host lattice. Therefore, we believe that the doped In_2O_3 film has more useful topographies for enhancing sensing performance.

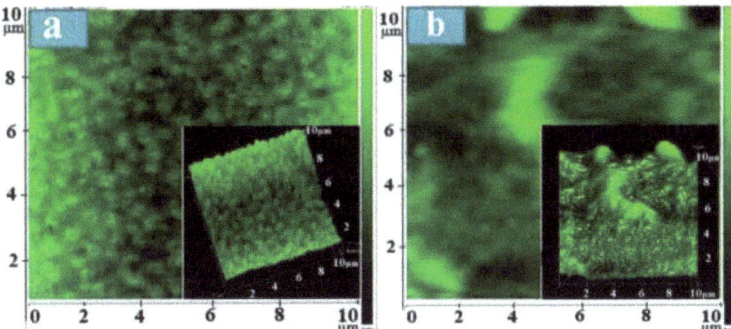

Figure 2. 2D and 3D AFM images of (**a**) pure In_2O_3 and (**b**) Cr doped In_2O_3 thin films.

3.3. Optical Analysis

Figure 3a shows the optical transmittance spectra of pure In_2O_3 and Cr doped In_2O_3 thin films. From the figure, it is clear that both the pure and doped In_2O_3 thin films are highly transparent in the visible region (~80%). The transmittance of Cr doped In_2O_3 film is lower when compared with the pure In_2O_3 film due the presence of Cr impurities. The decrease of transmittance might be due to increase in film thickness by external impurities. A steep absorption edge is originating near the UV region for both films; and, the edge is shifted to a higher wavelength side for the doped In_2O_3 film by absorbing the higher energy photons. This sharp fall near the UV range infers good crystalline nature of the wide bandgap of the prepared films. The optical band gap of the pure and Cr doped In_2O_3 films can be estimated from the relation [26].

$$\alpha h\upsilon = B\left(h\upsilon - E_g\right)^n \qquad (2)$$

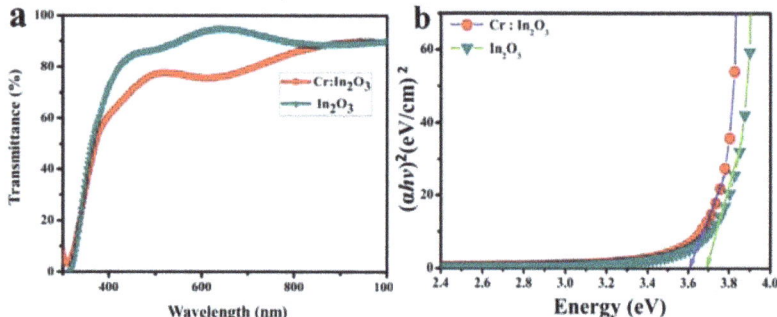

Figure 3. (a) Optical transmittance and (b) bandgap spectra of pure and Cr doped In$_2$O$_3$ films.

Optical bandgap of pure In$_2$O$_3$ and Cr doped In$_2$O$_3$ thin films are assumed from the plot of $(\alpha h\upsilon)^2$ vs. $h\upsilon$, as given in Figure 3b. The direct bandgap (E_g) values of pure and Cr doped In$_2$O$_3$ thin films were found to be decreased from 3.69 to 3.61 eV. Habib et al. [27] observed the decrease of bandgap from pure to Cr doped metal oxide thin films due to the formation of defects and/or the higher flow of electrons between valence band to the conduction band.

The observed lower bandgap value of Cr doped In$_2$O$_3$ thin film is attributed to the lower crystallite size obtained from XRD. In addition, the change of bandgap might be due to the influence of various factors such as particle size, doping impurities, carrier concentration, and deviation of elemental stoichiometry in the crystal lattice. In our case, there is a reduction of crystallite size, which causes the decrease of bandgap from pure to doped films.

3.4. X-ray Photoelectron Spectroscopy Analysis

Figure 4a shows the X-ray photoelectron survey scan spectrum of the Cr doped In$_2$O$_3$ thin film. The detailed survey scan spectrum has been performed to determine the ionic state of Indium (In), Chromium (Cr), and Oxygen (O) in the deposited Cr doped In$_2$O$_3$ thin film. From Figure 4a, there are several peaks observed related to the presence of different electronic states of In, Cr, and O.

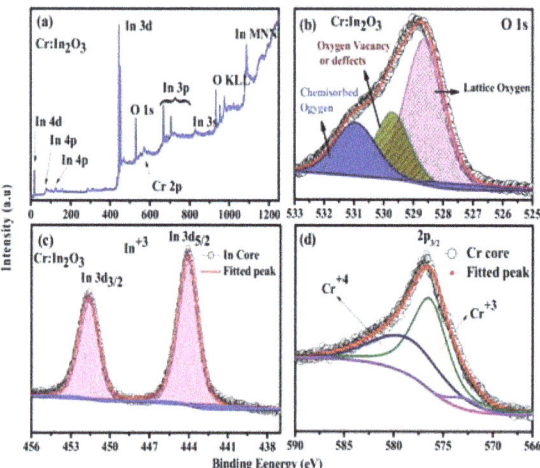

Figure 4. X-ray photoelectron spectra of (a) survey scan, (b) oxygen core level (c) indium core level and (d) chromium core level for the Cr doped In$_2$O$_3$ thin film.

Figure 4b shows the observed core level fine spectra of O 1s from the Cr doped In_2O_3 thin film. From Figure 4b, the deconvolution of the observed spectrum shows the presence of three distinguished peaks. The high intense peak exhibited at 528.4 eV related to the lattice oxygen, while the other higher binding energy peak at 531.2 eV is due to the presence of adsorbed oxygen on the film surface [21]. The centre peak located at 529.6 eV is due to the presence of oxygen vacancy in the film surface. Oxygen vacancy in the film is rather a familiar occurrence in transparent oxide films due to the loss of oxygen during the synthesis process. The presence of the oxygen vacancy state can be understood from the Cr core-level spectrum. In Figure 4c, it is observed that In (3d) core-level spectrum contains two peaks and the spin-orbit splitting is 7.6 eV in-between two peaks. Features of In-3d core-level spectrum are fitted with combined Gaussian function and the observed peaks corresponding to $In-3d_{3/2}$ and $In-3d_{5/2}$ states are observed at 451.6 and 444 eV, respectively [28]. These are consistent with the 3^+ state of indium (In) in In_2O_3 lattice. A similar spectrum was also observed for the pure In_2O_3 thin film.

Figure 4d shows the Cr $2p_{3/2}$ core level spectrum of Cr doped In_2O_3 thin film. The spectrum was corrected with Shirley background approximation and fitted with the combined Gaussian function. The fitted spectrum reflects the presence of Cr^{3+} state along with the Cr^{4+} state. The most intense peak at 576.3 eV is assigned as Cr^{3+} state and a hump at higher energy tail of Cr^{3+} state reflects the presence of higher oxidation state Cr^{4+}. The hump cannot be a satellite feature due to Cr^{3+} state since its position should appear at 11 eV higher in binding energy from the Cr $2p_{3/2}$ feature. A small feature of 572.7 eV is due to the multiple splitting of Cr 2p energy levels [29]. The occurrence of 3^+ and 4^+ states of Cr ions suggest its substitutional character and excludes the possibility of Cr clusters in the studied film. The hetero-valency of Cr ion can be explained by the occurrence of oxygen vacancy from the prepared film.

3.5. VSM Analysis

Figure 5 shows the magnetization vs. the applied magnetic field (M-H) curves of pure In_2O_3 and Cr doped In_2O_3 thin films measured at room temperature. The magnetization data of both the films were rectified by subtracting the effect of substrate contribution. From Figure 5, it is observed that the prepared films exhibit a tolerant ferromagnetism with an increase of saturation magnetization (M_s). It is a well-known fact that the pure In_2O_3 thin film exhibits ferromagnetic behaviour with lower M_s value when compared to a diamagnetic nature of stoichiometric bulk In_2O_3 [25]. This weak ferromagnetism of the In_2O_3 thin film might be due to the presence of some anion vacancies in the film [7]. From Figure 5, it is observed that Cr-doped In_2O_3 thin film shows higher saturation magnetization (M_s) value when compared to the pure In_2O_3 thin film due to the incorporation of Cr into the host lattice. The increase in saturation magnetic moment from pure In_2O_3 (23 emu/cc) to Cr doped In_2O_3 (39 emu/cc) thin films is at 10,000 (Oe). This may be due to the magnetic exchange between Cr^{3+} and In^{3+} around free electron trapped sites and also the oxygen vacancies along with trapped electrons, which leads to enhance the ferromagnetic behaviour. It is also clear that the Cr doped In_2O_3 thin film exhibits good ferromagnetism; though, the saturation magnetization value remains very close to pure In_2O_3. This fundamental ferromagnetism may be occurring from the magnetic exchange interactions of the trapped electrons in the anionic vacancies called as F-centred mediated ferromagnetism. This mechanism is extensively presumed to be the origin of ferromagnetism in metal oxide thin films as well as a DMS system. Therefore, we believe that such an F-centre mediated ferromagnetism occurs in our prepared film because of the vacuum deposited Cr doped In_2O_3 film, which leads to the creation of oxygen vacancies. Such oxygen vacancies can trap the free electrons, as described by Coey et al. [30], and these trapped electrons act as F-centres. These F-centres mediate and could lead to a magnetic exchange interaction process in the neighbouring metal cations.

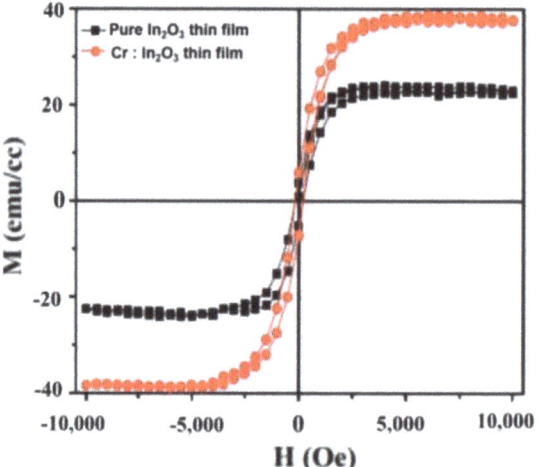

Figure 5. M-H curves of pure In$_2$O$_3$ and Cr doped In$_2$O$_3$ thin films measured at room temperature.

3.6. Ammonia (NH$_3$) Sensing Studies

The room temperature NH$_3$ sensing properties of the prepared pure In$_2$O$_3$ and Cr doped In$_2$O$_3$ films were studied using home-made gas sensing setup (Figure 6). The gas sensing response is mainly dependent on the surface morphology, operating temperature, and metal doping for metal oxide materials. Amongst them, the operating temperature plays a key role in improving the sensor's gas sensitivity because it controls the electrical resistivity by changing the activation energy. Han et al. [22] and Wang et al. [21] reported that the sensitivity of In$_2$O$_3$ is increased with the increasing operating temperature due to decrease in the activation energy and the molecular motion acceleration. However, scientists are focusing on improving its sensitivity at room temperature to ensure a safe and nontoxic environment in medical laboratories and chemical industries. [31]. In view of this, the performance of room temperature ammonia sensor for different NH$_3$ concentrations is reported in the present study.

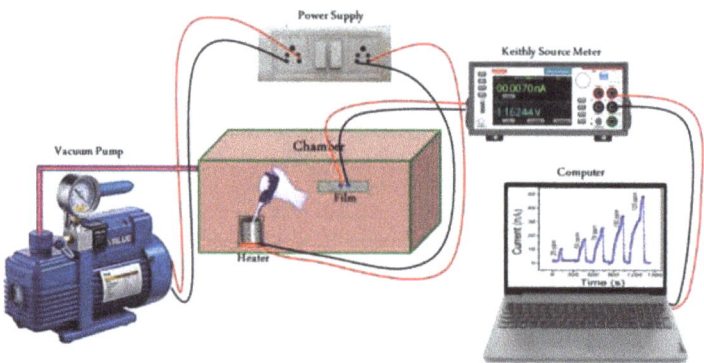

Figure 6. Schematic diagram of the ammonia gas sensing setup.

3.6.1. Sensitivity

The electrical current deviation of pure In$_2$O$_3$ and Cr doped In$_2$O$_3$ films were measured with different NH$_3$ concentrations, such as 25, 50, 75, 100, and 125 ppm, and are shown in Figure 7. It is observed that the base current (under air) value is slightly decreased from pure In_2O_3 (2.21 × 10^{-9} A) to Cr doped In_2O_3 film (1.01 × 10^{-9} A). This may possibly be due to suppression of the electron carrier concentration of host In_2O_3 lattice by the presence of surface defects of oxygen. Caricato et al. [32] observed the reduction of both mobility and carrier concentration when Cr metal doping into the ITO film. However, for the gas sensor applications, the surface related oxygen defects can enhance the sensing response by chemical reaction of both film surface and chemical species. As seen in Figure 7, the electrical current was significantly increased when increasing NH_3 concentrations for both pure and Cr doped In_2O_3 film sensors. This is because of more electrons moving to the conduction band once the NH_3 species interact with the surface of In_2O_3 film. When compared to pure In_2O_3, the current level reaches a maximum for Cr doped In_2O_3 film due to more surface defects by Cr inclusion.

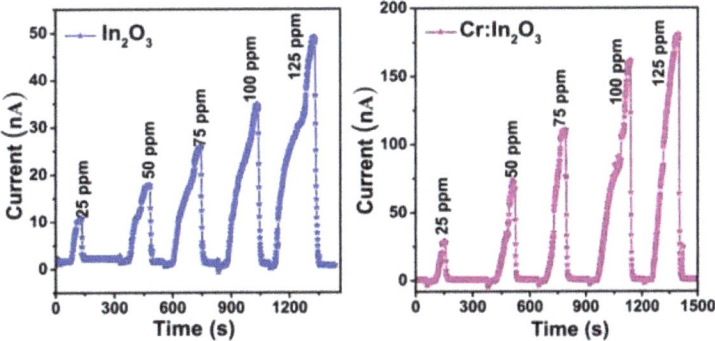

Figure 7. Electrical current variation of pure In_2O_3 and Cr doped In_2O_3 thin films at different NH_3 concentration (ppm).

From the observed current value under air and NH_3 atmosphere conditions, the gas sensitivity was calculated by using the following equation [31],

$$S = \frac{(I_g - I_a)}{I_a} \qquad (3)$$

where, I_g is the electrical current at gas atmosphere and I_a is the electrical current at air atmosphere. Figure 8 illustrates the sensitivity variation of pure and Cr doped In_2O_3 films with respect to NH_3 concentrations. It can be seen from the figure that the pure In_2O_3 film show a less sensitivity of 4% and then it reaches up to 28% (at 25 ppm of NH_3), while Cr is included in the In_2O_3 film. This can be explained on the change of surface activity and structural disorder by doping. Both XRD and AFM results evidenced that there is a decrease in the crystallite size and increase in the surface roughness for Cr doped In_2O_3 film compared to the pure In_2O_3. Due to having high surface roughness and/or large surface area, higher number of NH_3 molecules are adsorbed on the In_2O_3 film surface, which generates more oxygen ions, resulting in enhancement of the sensitivity. Moreover, it is noticed that the sensing response is steadily increased from 21% to 182% at 125 ppm of NH_3 for pure and Cr doped In_2O_3 films. The following reasons can be considered for the increase of sensitivity; (i) increase of the surface catalytic effect on the In_2O_3 surface caused by NH_3 adsorption (ii) the adsorbed oxygen ions heavily interact with more NH_3 species, (iii) the rate of combustion reaction between film surface and the testing gas is increased by increase in NH_3 concentration due to low activation energy. Hassan et al. [33] reported the increase of sensing response from the undoped ZnO to Cr doped ZnO thin films by the presence of oxygen defects in the host structure.

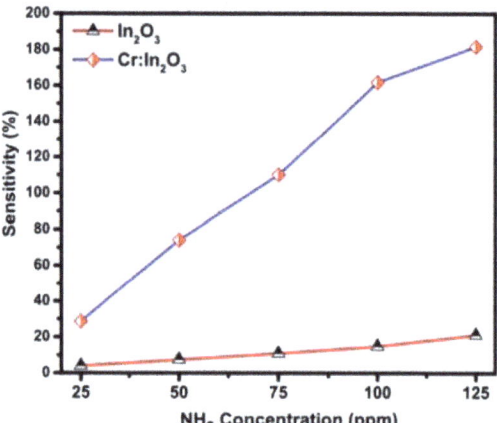

Figure 8. Variation of NH$_3$ sensitivity for pure In$_2$O$_3$ and Cr doped In$_2$O$_3$ thin films at different NH$_3$ concentrations.

3.6.2. Response and Recovery Speed

Response and recovery times are the key parameters for gas sensor applications. In general, the metal oxide-based sensor should be having a faster response and recovery time with respect to chemical species due to the adsorption of more oxygen ions on the film surface. The required time taken to reach 90% of the current is known as response time and the exact inverse trend is known as recovery time. Figure 9a illustrates the plot of response and recovery times for pure In$_2$O$_3$ thin film at 25 ppm of NH$_3$. The observed response and recovery times increase with increase of NH$_3$ concentration, as given in Figure 9b. This may be attributed to increasing the interaction rate of adsorbed oxygen and NH$_3$ species, and therefore, releasing electrons to the conduction band of host lattice. The obtained recovery time is less when compared to response time because the chemisorbed oxygen ions are taking lesser time to desorb from the film surface. In the present work, the observed response time is 28 s, and the corresponding recovery time is 6 s for Cr doped In$_2$O$_3$ film at 25 ppm. Hassan et al. [33] obtained the least response and recovery times, which are 15 and 72 s, respectively, for Cr doped ZnO thin film. Han et al. [22] reported response/recovery times of 14/10 s for Ce doped In$_2$O$_3$ nanospheres. In our case, both the values are nearly the same as the above reported values.

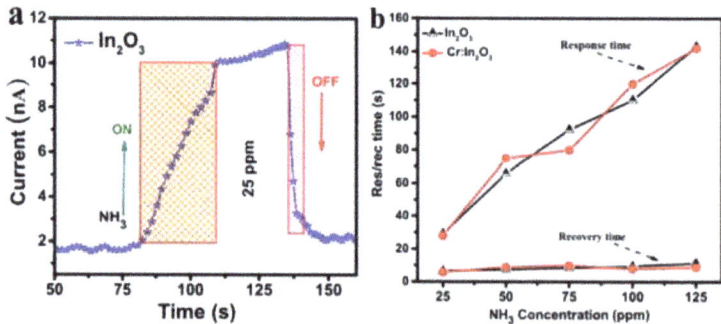

Figure 9. (a) Plot of response/recovery time at 25 ppm for pure In$_2$O$_3$, (b) variation of response and recovery times for pure and doped films with respect to different NH$_3$ concentrations.

3.6.3. Sensing Mechanism of NH_3/In_2O_3

In_2O_3 is one of the familiar semiconducting n-type sensing materials because of its surface-controlled type. The gas sensing performance of In_2O_3 sensor with respect to analyte gas can be explained by the changing in current upon exposure to different gas concentrations. Oxygen ions and its related defects play a key role in enhancing the sensing response, therefore, its necessary to understand both adsorption and desorption of oxygen ions under air and surrounding gas atmosphere. When the In_2O_3 film is exposed in air atmosphere, the electrical current is suppressed by the adsorbed O_2 molecules, which are converted as negative charged oxygen ions (O_2^-, O^- or O^{2-}) by extracting the electrons from the conduction band (CB). At room temperature, O_2^- ionized oxygen is formed in the prepared In_2O_3 film surface. Due to the creation of O_2^- ions, a depletion layer is formed. As a result, an initial low electrical current is observed. The formation of negatively charged O_2^- ions on the film surface can be written as $O_{2(ads)} + e \rightarrow O_2^-$ [34]. Under NH_3 atmosphere, the film surface reacts with chemical spices and releases the captured electrons back into the *CB*. Consequently, there is an increase in carrier concentrations. Therefore, the electron depletion layer becomes thin and can be described as $4NH_{3(ads)} + 2O^-_{(ads)} \rightarrow N_{2(ads)} + 6H_2O_{(ads)} + 2e^-$. The observed NH_3 performance can possibly be explained by the following reasons: (i) The higher surface roughness of Cr doped In_2O_3 provides fast chemisorb reaction and accelerates the negative charged ions, which leads to an increase in the sensing response, (ii) the incorporation of Cr^{3+} ions act as a donor by replacing In^{3+}, consequently, it can improve the oxygen vacancies, which leads to enhancement of sensitivity.

3.6.4. Relative Humidity (RH) and Repeatability

Relative humidity (RH) analysis is one of the major parameters to investigate the performance of NH_3 humidity for the pure and Cr doped In_2O_3 films at room temperature. Initially, In_2O_3 film was exposed to NH_3 at 125 ppm under various humidity conditions such as 33%, 54%, and 75%, which are found from the saturated solutions NaCl, $MgCl_2$, and $Mg(NO_3)_2$. Figure S2 (Supplementary File) shows the observed current variation with respect to time for both pure and Cr doped In_2O_3 films. Here, the saturated solutions NaCl, $MgCl_2$, Mg $(NO_3)_2$ are labelled as RH_1, RH_2, and RH_3, respectively, as given in Figure S2. Under humidity air condition, the H_2O molecules could be adsorbed on the In_2O_3 film surface, it converts a hydroxyl form and thus donates the electrons to the In_2O_3 lattice. Further, the adsorbed H_2O molecules can replace the oxygen ions in the film surface and create the electrons to the conduction band. Due to the above reasons, there is an increase of electron carrier concertation as well as electrical current as seen in Figure S2. The figure shows the reduction of electrical current on increasing relative humidity in percentage. This might be due to the H_2O molecules being directly chemisorbed on the oxygen vacancy sites [35]. Pandeeswari et al. [36] suggested the reduction of response being due to the hydrophobic nature of the film.

Repeatability of the pure In_2O_3 and Cr doped In_2O_3 films were accomplished for five consecutive cycles using 25 ppm of NH_3 are, which are given in Figure 10. The figure shows that the observed current level is maintaining the same range for all five cycles. Additionally, it is maintaining a fast response and recovery times towards 25 ppm of NH_3 species. Hence, we conclude that the prepared In_2O_3/In_2O_3:Cr films based metal oxide sensor has good repeatability and can be used in commercial sensors.

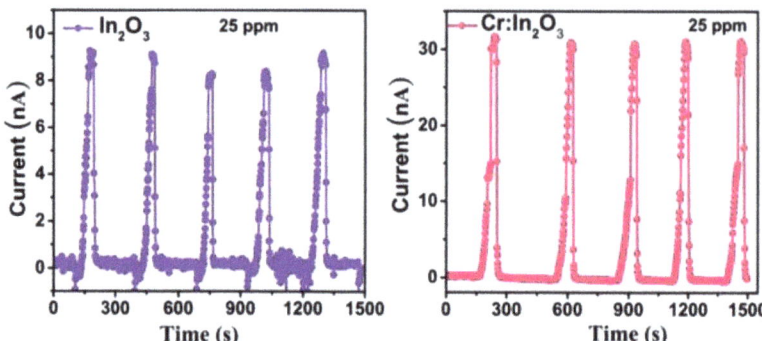

Figure 10. Repeatability for both pure In_2O_3 and Cr doped In_2O_3 thin films at 25 ppm of NH_3.

4. Conclusions

In summary, the successful fabrication of In_2O_3 and Cr-doped In_2O_3 thin films onto a Si substrate by PLD technique is reported in this article. The deposited films are crystalline and grown in a single phase without any impurity, which is confirmed by the XRD studies. The XPS analysis highlights that Cr doped In_2O_3 thin film has oxygen vacancies, which also lead to mixed oxidations states of Cr in 3^+ and 4^+ states. A decrease in the bandgap value from pure to doped In_2O_3 films is due to the existence of defects by Cr doping. The Cr doped In_2O_3 thin film sensor showed a maximum sensitivity of 182% at 125 ppm of NH_3 due to good material properties such as high surface roughness and defects of oxygen. Both AFM and XPS results support the enhancement of sensitivity from pure to Cr doped In_2O_3 thin film sensor. The doped In_2O_3 sensor also exhibits fast response and recovery times of 28 and 6 s respectively. The prepared In_2O_3/In_2O_3:Cr films have good repeatability; therefore, we propose that In_2O_3 based sensor could be used in commercial sensor devices.

Supplementary Materials: The following are available online at https://www.mdpi.com/article/10.3390/coatings11050588/s1, Figure S1: SEM micrographs of the (a) pure In_2O_3, (c) Cr doped In_2O_3 thin films; and EDX spectrum of (b) pure In_2O_3, (d) Cr doped In_2O_3 thin films, Figure S2: Relative humidity effect on the electric current for pure In_2O_3 and Cr doped In_2O_3 thin films.

Author Contributions: Conceptualization, Methodology of XRD, UV-Visible and XPS, writing of original draft, V.Y.; Conceptualization, methodology, data analysis and writing—original draft, H.P.K.; interpretation of data, writing—original draft, review and editing, K.D.A.K. and P.M.; discussion and reviewing of the manuscript, A.J.C. and K.V.G.; discussion and reviewing of the manuscript, S.A. and M.S.A.-B. All authors have read and agreed to the published version of the manuscript.

Funding: We would like to thank Taif University Research Supporting Project (number TURSP-2020/63), Taif University, Taif, Saudi Arabia.

Institutional Review Board Statement: Not applicable.

Informed Consent Statement: Not applicable.

Data Availability Statement: Data is contained within the article.

Acknowledgments: Authors thank (acknowledge) the director UGC DAE Consortium for Scientific Research Indore for providing experimental facilities for PLD, GIXRD, and XPS.

Conflicts of Interest: The authors declare no conflict of interest.

References

1. Bader, S.D.; Parkin, S.S.P. Spintronics. *Annu. Rev. Condens. Matter Phys.* **2010**, *1*, 71–88. [CrossRef]
2. Datta, S.; Das, B. Electronic analog of the electro-optic modulator. *Appl. Phys. Lett.* **1990**, *56*, 665–667. [CrossRef]
3. Miwa, S.; Ishibashi, S.; Tomita, H.; Nozaki, T.; Tamura, E.; Ando, K.; Mizuochi, N.; Saruya, T.; Kubota, H.; Yakushiji, K.; et al. Highly sensitive nanoscale spin-torque diode. *Nat. Mater.* **2013**, *13*, 50–56. [CrossRef] [PubMed]

4. Žutić, I.; Fabian, J.; Das Sarma, S. Spintronics: Fundamentals and applications. *Rev. Mod. Phys.* **2004**, *76*, 323–410. [CrossRef]
5. Ertler, C.; Fabian, J. Resonant tunneling magnetoresistance in coupled quantum wells. *Appl. Phys. Lett.* **2006**, *89*, 242101. [CrossRef]
6. Yoo, J.W.; Chen, C.Y.; Jang, H.W.; Bark, C.W.; Prigodin, V.N.; Eom, C.B.; Epsteinn, A.J. Erratum: Spin injection/detection using an organic-based magnetic semiconductor. *Nat. Mater.* **2010**, *9*, 778. [CrossRef]
7. Watanabe, S.; Ando, K.; Kang, K.; Mooser, S.; Vaynzof, Y.; Kurebayashi, H.; Saitoh, E.; Sirringhaus, H. Polaron spin current transport in organic semiconductors. *Nat. Phys.* **2014**, *10*, 308–313. [CrossRef]
8. Boales, J.A.; Boone, C.T.; Mohanty, P. Nanomechanical detection of the spin Hall effect. *Phys. Rev. B* **2016**, *93*, 161414. [CrossRef]
9. Hamberg, I.; Granqvist, C.G. Evaporated Sn-doped In_2O_3 films: Basic optical properties and applications to energy-efficient windows. *J. Appl. Phys.* **1986**, *60*, R123–R160. [CrossRef]
10. Odaka, H.; Iwata, S.; Taga, N.; Ohnishi, S.; Kaneta, Y.; Shigesato, Y. Study on electronic structure and optoelectronic properties of indium oxide by first-principles calculations. *Jpn. J. Appl. Phys.* **1997**, *36*, 5551–5554. [CrossRef]
11. Ginley, D.S.; Bright, C. Transparent conducting oxides. *MRS Bull.* **2000**, *25*, 15–18. [CrossRef]
12. Kumar, K.D.A.; Mele, P.; Ponraj, J.S.; Haunsbhavi, K.; Varadharajaperumal, S.; Alagarasan, D.; Algarni, H.; Angadi, B.; Murahari, P.; Ramesh, K. Methanol solvent effect on photosensing performance of AZO thin films grown by nebulizer spray pyrolysis. *Semicond. Sci. Technol.* **2020**, *35*, 085013. [CrossRef]
13. Batzill, M.; Diebold, U. The surface and materials science of tin oxide. *Prog. Surf. Sci.* **2005**, *79*, 47–154. [CrossRef]
14. Philip, J.R.; Punnoose, A.; Kim, B.I.; Reddy, K.M.; Layne, S.P.; Holmes, J.; Satpati, B.; LeClair, P.; Santos, T.S.; Moodera, J.S. Carrier-controlled ferromagnetism in transparent oxide semiconductors. *Nat. Mater.* **2006**, *5*, 298–304. [CrossRef] [PubMed]
15. Jayakumar, O.D.; Gopalakrishnan, I.K.; Kulshreshtha, S.K.; Gupta, A.; Rao, K.V.; Louzguine-Luzgin, D.V.; Inoue, A.; Glans, P.-A.; Guo, J.; Samanta, K.; et al. Structural and magnetic properties of $(In_{1−x}Fe_x)_2O_3$ ($0.0 \leq x \leq 0.25$) system: Prepared by gel combustion method. *Appl. Phys. Lett.* **2007**, *91*, 052504. [CrossRef]
16. Garcia, M.A.; Ruiz-González, M.L.; Quesada, A.; Costa-Kramer, J.L.; Fernandez, J.F.; Khatib, S.J.; Wennberg, A.; Caballero, A.C.; Martín-González, M.S.; Villegas, M.; et al. Interface Double-Exchange Ferromagnetism in the Mn-Zn-O System: New Class of Biphase Magnetism. *Phys. Rev. Lett.* **2005**, *94*, 217206. [CrossRef] [PubMed]
17. Chang-Yup, P.; Chun-Yeol, Y.; Kun-Rok, J.; Sung, C.S. Charge-carrier mediated ferromagnetism in Mo-doped In_2O_3 films. *Appl. Phys. Lett.* **2012**, *100*, 222409.
18. Kharel, P.; Sudakar, C.; Sahana, M.B.; Lawes, G.; Suryanarayanan, R.; Naik, R.R.; Naik, V.M. Room temperature ferromagnetism in Cr-doped In_2O_3 on high vacuum annealing of thin films and bulk samples. *J. Appl. Phys.* **2007**, *101*, 09H117. [CrossRef]
19. Ukah, N.; Gupta, R.; Kahol, P.; Ghosh, K. Influence of oxygen growth pressure on laser ablated Cr-doped In_2O_3 thin films. *Appl. Surf. Sci.* **2009**, *255*, 9420–9424. [CrossRef]
20. Xu, J.; Han, J.; Zhang, Y.; Sun, Y.; Xie, B. Studies on alcohol sensing mechanism of ZnO based gas sensors. *Sens. Actuators B Chem.* **2008**, *132*, 334–339. [CrossRef]
21. Wang, Z.; Hou, C.; De, Q.; Gu, F.; Han, D. One-step synthesis of Co-doped In_2O_3 nanorods for high response of formaldehyde sensor at low temperature. *ACS Sens.* **2018**, *3*, 468–475. [CrossRef] [PubMed]
22. Han, D.; Song, P.; Zhang, S.; Zhang, H.; Xu, Q.; Wang, Q. Enhanced methanol gas-sensing performance of Ce-doped In_2O_3 porous nanospheres prepared by hydrothermal method. *Sens. Actuators B Chem.* **2015**, *216*, 488–496. [CrossRef]
23. Manivasaham, A.; Ravichandran, K.; Subha, K. Light intensity effects on the sensitivity of ZnO:Cr gas sensor. *Surf. Eng.* **2017**, *33*, 866–876. [CrossRef]
24. Kumar, K.D.A.; Valanarasu, S.; Ganesh, V.; Shkir, M.; Kathalingam, A.; AlFaify, S. Effect of precursors on key opto-electrical properties of successive ion layer adsorption and reaction-prepared Al:ZnO thin films. *J. Electron. Mater.* **2018**, *47*, 1335–1343. [CrossRef]
25. Krishna, N.S.; Kaleemulla, S.; Amarendra, G.; Rao, N.M.; Krishnamoorthi, C.; Omkaram, I.; Reddy, D.S. Structural, optical and magnetic properties of Cr doped In_2O_3 powders and thin films. *J. Mater. Sci. Mater. Electron.* **2015**, *26*, 8635–8643. [CrossRef]
26. Anand, V.; Sakthivelu, A.; Kumar, K.D.A.; Valanarasu, S.; Ganesh, V.; Shkir, M.; AlFaify, S.; Algarni, H. Rare earth Eu^{3+} co-doped AZO thin films prepared by nebulizer spray pyrolysis technique for optoelectronics. *J. Sol-Gel Sci. Technol.* **2018**, *86*, 293–304. [CrossRef]
27. Habib, I.Y.; Tajuddin, A.A.; Noor, H.A.; Lim, C.M.; Mahadi, A.H.; Kumara, N.T.R.N. Enhanced Carbon monoxide-sensing properties of Chromium-doped ZnO nanostructures. *Sci. Rep.* **2019**, *9*, 1–12. [CrossRef]
28. Beena, D.; Lethy, K.; Vinodkumar, R.; Pillai, V.M.; Ganesan, V.; Phase, D.; Sudheer, S. Effect of substrate temperature on structural, optical and electrical properties of pulsed laser ablated nanostructured indium oxide films. *Appl. Surf. Sci.* **2009**, *255*, 8334–8342. [CrossRef]
29. Tripathi, M.; Choudhary, R.J.; Phase, D.M. Phase coexistence and the magnetic glass-like phase associated with the Morin type spin reorientation phase transition in $SmCrO_3$. *RSC Adv.* **2016**, *6*, 90255–90262. [CrossRef]
30. Coey, J.M.D.; Venkatesan, M.; Fitzgerald, C.B. Donor impurity band exchange in dilute ferromagnetic oxides. *Nat. Mater.* **2005**, *4*, 173–179. [CrossRef]
31. Kumar, K.D.A.; Valanarasu, S.; Ponraj, J.S.; Fernandes, B.J.; Shkir, M.; AlFaify, S.; Murahari, P.; Ramesh, K. Effect of Er doping on the ammonia sensing properties of ZnO thin films prepared by a nebulizer spray technique. *J. Phys. Chem. Solids* **2020**, *144*, 109513. [CrossRef]

32. Caricato, A.P.; Cesaria, M.; Luches, A.; Martino, M.; Maruccio, G.; Valerini, D.; Catalano, M.; Cola, A.; Manera, M.G.; Lomascolo, M.; et al. Electrical and optical properties of ITO and ITO/Cr-doped ITO films. *Appl. Phys. A* **2010**, *101*, 753–758. [CrossRef]
33. Hassan, M.M.; Khan, W.; Mishra, P.; Islam, S.; Naqvi, A. Enhancement in alcohol vapor sensitivity of Cr doped ZnO gas sensor. *Mater. Res. Bull.* **2017**, *93*, 391–400. [CrossRef]
34. Xia, H.; Wang, Y.; Kong, F.; Wang, S.; Zhu, B.; Guo, X.; Zhang, J.; Wu, S. Au-doped WO_3-based sensor for NO_2 detection at low operating temperature. *Sens. Actuators B Chem.* **2008**, *134*, 133–139. [CrossRef]
35. Santhosam, A.J.; Ravichandran, K.; Ahamad, T. Donated free electrons induced enhancement in the NH_3 sensing ability of ZnO thin films—Effect of terbium loading. *Sens. Actuators A Phys.* **2020**, *316*, 112376. [CrossRef]
36. Pandeeswari, R.; Jeyaprakash, B. High sensing response of β-Ga2O3 thin film towards ammonia vapours: Influencing factors at room temperature. *Sens. Actuators B Chem.* **2014**, *195*, 206–214. [CrossRef]

Erratum

Erratum: Yaragani et al. Structural, Magnetic and Gas Sensing Activity of Pure and Cr Doped In$_2$O$_3$ Thin Films Grown by Pulsed Laser Deposition. *Coatings* 2021, *11*, 588

Veeraswamy Yaragani [1,*,†], Hari Prasad Kamatam [1,2,†], Karuppiah Deva Arun Kumar [3], Paolo Mele [3,*], Arulanandam Jegatha Christy [4], Kugalur Venkidusamy Gunavathy [5], Sultan Alomairy [6] and Mohammed Sultan Al-Buriahi [7]

1. Department of Physics, Institute of Aeronautical Engineering, Dundigal, Hyderabad 500043, India; k.hariprasad500@gmail.com
2. Department of Physics, Pondicherry University, Puducherry 605014, India
3. College of Engineering, Shibaura Institute of Technology, Saitama 337-8570, Japan; i048267@shibaura-it.ac.jp
4. Department of Physics, Jayaraj Annapackiam College for Women, Periyakulam, Theni 625601, India; jegathachristy@gmail.com
5. Department of Physics, Kongu Engineering College, Perundurai 638060, India; gunavathy@kongu.ac.in
6. Department of Physics, College of Science, Taif University, P.O. Box 11099, Taif 21944, Saudi Arabia; mohsaa996@gmail.com
7. Department of Physics, Sakarya University, Sakarya 54100, Turkey; mohammed.al-buriahi@ogr.sakarya.edu.tr
* Correspondence: veerayaragani@gmail.com (V.Y.); pmele@shibaura-it.ac.jp (P.M.)
† Author contributed equally.

The authors wish to make the following changes to their published paper [1]. On page 1, the first and co-corresponding author has been included: First author with co-corresponding author: Veeraswamy Yaragani (Institute of Aeronautical Engineering, Dundigal, Hyderabad-500 043, India).

On page 12, authors thank (acknowledge) the director UGC DAE Consortium for Scientific Research Indore for providing experimental facilities for PLD, GIXRD, and XPS. The funding source remains available in the published version.

The authors apologize for any inconvenience caused and state that the scientific conclusions are unaffected. The original article has been updated.

Reference

1. Yaragani, V.; Kamatam, H.P.; Deva Arun Kumar, K.; Mele, P.; Christy, A.J.; Gunavathy, K.V.; Alomairy, S.; Al-Buriahi, M.S. Structural, Magnetic and Gas Sensing Activity of Pure and Cr Doped In$_2$O$_3$ Thin Films Grown by Pulsed Laser Deposition. *Coatings* 2021, *11*, 588. [CrossRef]

Article

Enhanced Electrical Properties and Stability of P-Type Conduction in ZnO Transparent Semiconductor Thin Films by Co-Doping Ga and N

Chien-Yie Tsay * and Wan-Yu Chiu

Department of Materials Science and Engineering, Feng Chia University, Taichung 40724, Taiwan; Chiuwanyu@gmail.com
* Correspondence: cytsay@mail.fcu.edu.tw; Tel.: +886-4-2451-7250 (ext. 5312)

Received: 20 October 2020; Accepted: 2 November 2020; Published: 6 November 2020

Abstract: P-type ZnO transparent semiconductor thin films were prepared on glass substrates by the sol-gel spin-coating process with N doping and Ga–N co-doping. Comparative studies of the microstructural features, optical properties, and electrical characteristics of ZnO, N-doped ZnO (ZnO:N), and Ga–N co-doped ZnO (ZnO:Ga–N) thin films are reported in this paper. Each as-coated sol-gel film was preheated at 300 °C for 10 min in air and then annealed at 500 °C for 1 h in oxygen ambient. X-ray diffraction (XRD) examination confirmed that these ZnO-based thin films had a polycrystalline nature and an entirely wurtzite structure. The incorporation of N and Ga–N into ZnO thin films obviously refined the microstructures, reduced surface roughness, and enhanced the transparency in the visible range. X-ray photoelectron spectroscopy (XPS) analysis confirmed the incorporation of N and Ga–N into the ZnO:N and ZnO:Ga–N thin films, respectively. The room temperature PL spectra exhibited a prominent peak and a broad band, which corresponded to the near-band edge emission and deep-level emission. Hall measurement revealed that the ZnO semiconductor thin films were converted from n-type to p-type after incorporation of N into ZnO nanocrystals, and they had a mean hole concentration of 1.83×10^{15} cm^{-3} and a mean resistivity of 385.4 $\Omega\cdot$cm. In addition, the Ga–N co-doped ZnO thin film showed good p-type conductivity with a hole concentration approaching 4.0×10^{17} cm^{-3} and a low resistivity of 5.09 $\Omega\cdot$cm. The Ga–N co-doped thin films showed relatively stable p-type conduction (>three weeks) compared with the N-doped thin films.

Keywords: p-type oxide semiconductor; ZnO thin film; N doping; Ga–N co-doping; sol-gel spin coating; electrical stability

1. Introduction

Wide-bandgap oxide semiconductors have gained considerable attention and received great interest due to their potential for applications in optoelectronic devices, photovoltaic devices, and transparent oxide electronics, such as ultraviolet photodetectors, short wavelength light-emitting diodes, thin-film solar cells, transparent resistive random access memory, and transparent field-effect transistors [1–4]. The fabrication of light-emitting diodes, p-n junction photodetectors, and complementary metal-oxide-semiconductor devices requires the combination of both n-type and p-type oxide semiconductors. The lack of availability of stable p-type oxide semiconductors limits the electrical efficiency of p-n junction-based devices or bipolar device applications [3,5]. Therefore, the development of stable and reproducible p-type oxide semiconductor thin films is necessary.

The crystalline wurtzite ZnO exhibits n-type conductivity due to a deviation from chemical stoichiometry and the presence of native point defects such as O vacancies (Vo) and Zn interstitials (Zn_i)

that makes the electrons easily be excited to the conduction band [2,6]. The difficulty in achieving p-type ZnO semiconductors is caused by a combination of deep acceptor levels, the low solubility of acceptors, and a strong self-compensation effect [7,8]. Previous studies demonstrated that the impurity doping of ZnO is an important approach for modulating the carrier concentration and electrical resistivity to satisfy device applications [9,10]. A sufficient amount of acceptors should be generated to overcome the n-type back ground, such as any donor-like defects created through preparation processes. It has been suggested that the group V element nitrogen (N) may be a potential acceptor dopant for O site substitution in ZnO to obtain p-type conductivity, for N has the smallest ionization energy among the group V dopants (N, P, and As) [2,11]. In addition, N (0.146 nm) and O (0.140 nm) have close ionic radii, and so N ions may easily be substituted on O sites of the ZnO lattice [12]. Moreover, the N–O bond length (1.88 Å) is close to the Zn–O bond length (1.93 Å) [13]. Kim's group reported that the incorporation of N as a dopant changed the conductivity of sol-gel-derived ZnO thin films from n-type to p-type and produced ZnO:N thin films with a hole concentration of 8.11×10^{16} cm^{-3} when the N doping content reached 20 at.% in the resultant solution [14]. However, the production of device-quality p-type conduction in N-doped ZnO thin films remains a challenge due to the low dopant solubility in the host material and high ionization energy [11].

Co-doping of two kinds of extrinsic atoms (using acceptors and donors simultaneously) is considered an effective method for producing stable p-type conductivity in ZnO semiconductor thin films [7,15,16]. Previous experimental studies found that acceptor and donor co-doped ZnO thin films not only exhibited p-type conductivity but also showed reasonable stability and reproducibility in electrical characteristics [17–19]. It was reported that simultaneous doping of nitrogen and group III elements can improve the hole concentration and enhance the stability of p-type ZnO thin film because the doping of the reactive donor element increases the solubility of nitrogen in ZnO [20–23]. According to first-principles calculations, Yan et al. proposed that the formation of passive acceptor-donor (Ga–N) complexes can create an impurity band above the valence-band maximum (VBM) in ZnO crystal based on reduction of the ionization energies of dopants [8]. A theoretical study by Duan et al. reported that the ionization energy of Ga–2N complex is lower than that of the single N acceptor [11]. The generation of N–Ga–N complexes could increase the solubility percentage of N ionized dopants and achieve shallower acceptor levels [6]. A previous report noted that the co-doping of Ga and N in ZnO thin films achieved the best p-type character among the types of impurity co-doped ZnO thin films [8]. Theoretical calculations also predict that Ga–N co-doped ZnO would yield better p-type conduction than Al–N co-doped ZnO [11]. This prediction is consistent with numerous experimental finding.

Several thin film deposition techniques have been employed to produce the group III (B, Al, Ga, and In)–N co-doped p-type ZnO semiconductor thin films including chemical vapor deposition, molecular-beam epitaxy, pulsed laser deposition, radio-frequency (rf) magnetron sputtering, spray pyrolysis, and the sol-gel method [2,7]. Most researchers have focused on the preparation of p-type ZnO thin films and investigated the effects of acceptor-donor co-doping on the physical properties. The authors noted that few studies have reported the electrical stability of solution-processed p-type ZnO thin films. There is no report on Ga–N co-doping p-type ZnO thin films prepared by the sol-gel method. Therefore, we prepared ZnO, ZnO:N, and ZnO:Ga–N thin films on glass substrates via the sol-gel method and spin-coating technique. The structural-physical properties relation of ZnO-based semiconductor thin films was investigated in detail through variation of the structural features to examine if they affect the optical properties and electrical characteristics. In addition, the electrical stability of p-type ZnO-based thin films was verified and is reported in this study.

2. Materials and Methods

2.1. Coating Solution Synthesis and Thin Film Deposition

The series of ZnO-based thin films, including undoped ZnO, N-doped ZnO (ZnO:N), and N–Ga co-doped ZnO (ZnO:N–Ga), were deposited on alkali-free glass substrates (Nippon Electric Glass

OA-10, 50 × 50 × 0.7 mm³ in size) using the sol-gel method and spin coating technique. Prior to the deposition of the thin films, the glass substrates were ultrasonically cleaned in detergent, isopropyl alcohol, and acetone, washed thoroughly with distilled water and dried on a hot plate at 110 °C for 15 min. Three kinds of analytical reagent (AR)-grade metallic salts, including zinc acetate dihydrate ($Zn(CH_3CO_2)_2 \cdot 2H_2O$, J.T. Baker, Phillipsburg, NJ, USA), ammonium acetate (CH_3COONH_4, Sigma, St. Louis, MO, USA), and gallium (III) nitrate hydrate ($Ga(NO_3)_3 \cdot xH_2O$, Alderich, St. Louis, MO, USA), were chosen as the source materials of zinc, nitrogen, and gallium, respectively. The first precursor solutions were synthesized by dissolving a stoichiometric ratio of AR-grade metallic salts in 2-methoxyethanol (solvent) and diethanolamine (stabilizing agent) using a stirring hot plate (PC-420D, Corning, Glendale, AZ, USA) at 60 °C for 2 h until the resultant solution became clear and transparent. After that procedure, the synthesized sols were aged for 72 h at room temperature under atmospheric environment before being used in the spin-coating process. The concentration of the amount of metal ions in the resultant solutions was maintained at 0.4 M [24] and the molar ratio of metal ions to stabilizing agent was kept at 1:1. In addition, the atomic ratio of [N]/[Zn + N] was 0.3 for two impurity doped thin films (ZnO:N and ZnO:N–Ga), and that of [Ga]/[Zn + N] was 0.01 for ZnO:N–Ga thin films. The respective sols were spin coated on the pre-cleaned glass substrates at 1000 rpm for 30 s, followed by preheating at 300 °C in air for 10 min to evaporate volatile materials. That procedure was repeated three times, and then the dried sol-gel films were annealed at 500 °C in oxygen ambient for 1 h to remove residual organic compounds of the gel films and achieve crystalline oxide thin films [22].

2.2. Characterization Techniques

The crystalline structure and crystallinity of the as-prepared ZnO-based thin films were analyzed by Bruker D8 Discover SSS high-resolution X-ray diffractometer (XRD, Bruker, Karlsruhe, Germany) with Cu-Kα radiation (λ = 1.5418 Å) by glancing incidence technique at an incident angle of 0.8°. Plane-view and cross-sectional view micrographs of the thin film samples were acquired by Hitachi S-4800 field-emission scanning electron microscope (FE-SEM, Hitachi, Tokyo, Japan). A Digital Instrument scanning probe microscope (SPM, NS4/D3100CL/MultiMode, Digital Instrument, Mannheim, Germany) was employed for characterizing the surface morphology and surface roughness of oxide films in tapping mode with a tip scanning rate of 0.7 Hz. The chemical bonding energies and elemental composition of the oxide films were determined by X-ray photoelectron spectroscopy (XPS, ULVAC-PHI PHI 5000 VersaProbe, Kanagawa, Japan). Prior to the XPS examination, Ar ion sputter-etching was performed for 15 s to remove the surface contamination. The optical transmission characteristics of the samples of glass/ZnO-based thin films were examined with a Hitachi U-2900 ultraviolet-visible (UV-Vis) spectrophotometer (Hitachi, Tokyo, Japan). Room-temperature (RT) emission spectra of three ZnO-based thin films were measured by the HORIBA Jobin Yvon photoluminescence (PL) spectrometer (LabRAM HR, Paris, France) with an excitation wavelength of 325 nm. The electrical characteristics of each thin film sample were measured by the van der Pauw method using the Hall-effect measurement system (HMS-3000, Ecopia, Gyeonggi-do, Korea) with a magnetic field of 0.55 T at RT.

3. Results and Discussion

Various types of nitrogen (N) sources, including NH_3, N_2, NO, N_2O, and CH_3COONH_4 have been successfully used to prepare of device-quality p-type ZnO semiconductor thin films. In this study, the N dopant was obtained through decomposition of ammonium acetate to acetamide and then to CO, NO, and NO_2 [24]. The NO and NO_2 acted as the N sources. It was suggested that NO easily forms N_O acceptors, which are shallow acceptors in ZnO [2,8]. Moreover, it was necessary to provide oxygen to suppress oxygen vacancy (Vo) formation during the preparation of solution-processed p-type ZnO thin films. Because Vo may form as a compensation center in p-type ZnO [25], we performed the post deposition crystallization heat treatment in pure oxygen atmosphere to decrease the background carrier concentration.

The results of the XRD examination, shown in Figure 1, indicate that the three kinds of ZnO-based thin films were polycrystalline in nature. The intense sharp diffraction peaks suggested that the pure ZnO thin film had better crystallinity (pattern (i) in Figure 1) than those of the two impurity-doped ZnO thin films. Previous studies reported that the crystalline quality of oxide thin films would be degraded by impurity doping [10,26,27]. Comparison of collected XRD patterns (in 2θ range from 25° to 65°) with the standard JCPDS file of ZnO crystal (JCPDS No. 36-1451) indicated the hexagonal wurtzite phase in the ZnO-based thin films. In addition, no X-ray diffraction signals related to other crystal phases, such as GaN, Ga_2O_3, and Zn_3N_2, were detected from the set of thin film samples, indicating the formation of a single phase.

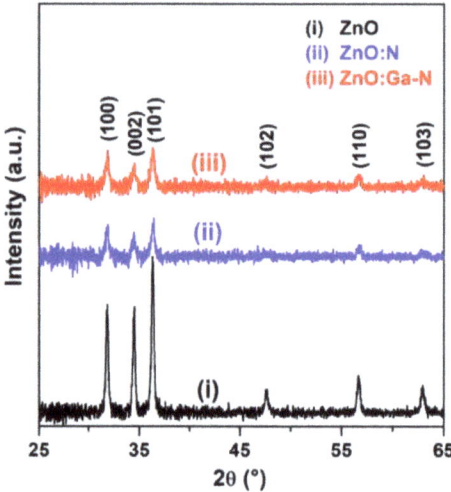

Figure 1. X-ray diffraction (XRD) patterns of ZnO, N-doped ZnO (ZnO:N), and Ga–N co-doped ZnO (ZnO:Ga–N) thin films deposited on glass substrates by the sol-gel spin-coating process.

A preferential orientation may occur due to nucleation, horizontal growth of the nuclei and vertical growth of the closed layer. As shown in Figure 1, we not only confirmed that these as-prepared ZnO thin films had a predominance of hexagonal wurtzite phase but also observed that they achieved a favorable orientation along the (101) plane. The relative intensities of $I_{(101)}/I_{(1000) + (002) + (101)}$ for the ZnO, ZnO:N, and ZnO:Ga–N samples were 0.422, 0.406, and 0.387, respectively. The calculated results reveal that the relative diffraction intensity of the undoped sample (0.422) was significantly smaller than that of the relative intensity (0.497) obtained from the standard JCPDS database; incorporation of N and Ga–N into the ZnO changed the growth rate of the (101) plane such that the three main diffraction planes had close growth rates. It is noted that the diffraction signals of the (102) and (103) planes were very weak for the two impurity-doped samples, implying an apparent decrease in crystallinity. According to the above discussion, we found that these sol-gel-derived ZnO-based thin films had no preferential orientation; impurity doping favored random growth of crystallites and caused degradation of the crystallinity. The average crystallite sizes of the thin film samples were quantified from the full widths at half-maximum and the Bragg diffraction angles of the three major diffraction peaks, (100), (002), and (101), with Scherrer's formula and are presented in Table 1. Table 1 summarizes our examined and calculated results on the structural features and optical properties. The average crystallite sizes for the undoped, N doped and Ga–N co-doped ZnO thin films were 26.9, 20.4, and 17.0 nm, respectively. It was determined that all the obtained ZnO-based thin films consisted of nanocrystals; N doping significantly inhibited the crystal growth rate due to lattice distortion, and the Zn:Ga–N samples had the smallest average crystallite size.

Table 1. Microstructural features and optical parameters of the obtained ZnO-based thin films.

Sample	Average Crystallite Size (nm)	Root Mean Square (RMS) Roughness (nm)	Average Transmittance [a] (%)	Absorption Reflectance [b] (cm^{-1})	Optical Bandgap (eV)	Urbach Energy (meV)
ZnO	26.9	3.6	85.14	19,019	3.27	89.2
ZnO:N	20.4	2.7	89.77	24,020	3.24	123.8
ZnO:Ga–N	17.0	1.1	90.29	25,574	3.24	148.5

The average transmittance values [a] and average absorption coefficients [b] were calculated from the recorded transmittance data of wavelengths from 400 to 800 nm.

Figure 2 shows the cross-sectional FE-SEM micrographs of ZnO-based thin films deposited onto glass substrates, which revealed good thickness uniformity. It can be seen that these oxide thin films possessed significantly granular microstructures, had homogeneous particle size distributions, and exhibited nano-sized pores inside the sol-gel-derived ZnO-based thin films. In addition, it was significantly observed that the mean particle size of the N-doped thin film was smaller than that of the undoped thin film, and the Ga–N co-doped thin film possessed the finest grade of particles. The microstructural features of the developed thin films, observed by FE-SEM, were consistent with the results of the XRD examination. The thickness of each thin film sample was estimated from the corresponding cross-sectional FE-SEM micrograph. The mean film thicknesses of the undoped, N-doped, and Ga–N co-doped ZnO thin films were found to be 85, 45, and 40 nm, respectively. The difference in film thickness between the three kinds of ZnO-based thin films can be ascribed to the variation in the viscosities of the coating solutions with the concentration of solute (metallic salts). The corresponding plane-view FE-SEM micrographs are shown in Figure S1. In that figure, it is clearly shown that the free surfaces of the films were flat, with no visible pores or cracks in the observation area.

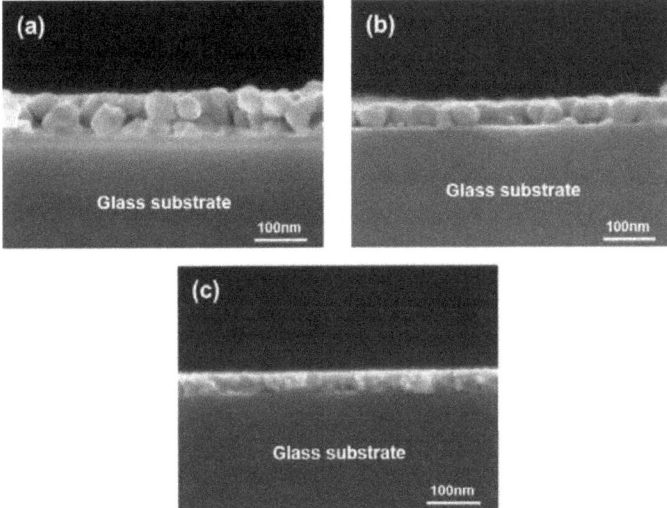

Figure 2. Cross-sectional field-emission scanning electron microscope (FE-SEM) micrographs of ZnO-based thin films: (**a**) undoped, (**b**) N-doped, and (**c**) Ga–N co-doped samples.

The topographic features of each thin film sample were examined by SPM to study the surface morphology and measure the root-mean-square (RMS) roughness. SPM images of the sol-gel-derived ZnO-based thin films deposited on glass substrates are presented in Figure 3. Those images show a uniformly granular morphology and dense surface. The significant differences between the three SPM images were the particle size and surface flatness. It was observed that the co-doped sample had the smallest particles and exhibited a flat surface morphology. The average particle sizes of the ZnO, ZnO:N, and ZnO:Ga–N thin films, as estimated from the corresponding 2D SPM images (Figure S2) in

ImageJ software, were 77.3, 46.2, and 39.3 nm, respectively. The observation of surface morphological characteristics of the thin film samples by SPM had a good agreement with the XRD results. Table 1 lists the values of RMS roughness of the ZnO, ZnO:N, and ZnO:Ga–N thin film samples, which were 3.6, 2.7, and 1.1 nm, respectively. As noted above, the impurity doping reduced the particle (crystallite) size and thereby decreased the surface roughness.

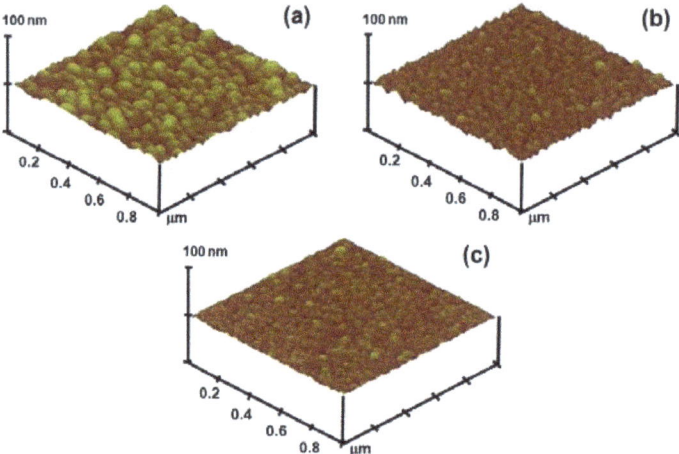

Figure 3. Three-dimensional (3D) scanning probe microscope (SPM) micrographs of the free surfaces of ZnO-based thin films: (**a**) undoped, (**b**) N-doped, and (**c**) Ga–N co-doped samples.

XPS examinations were performed to determine the elements responsible and identify the chemical bonding states of the ZnO-based thin films produced in this study. The narrow scan XPS spectra of Zn, O, and N regions of ZnO-based thin films are presented in Figure 4. The core level peaks of Zn 2p are presented in Figure 4a; they revealed similar XPS spectra for the three samples. The peaks of binding energies at 1021.0 and 1044.2 eV for the ZnO sample corresponded to the Zn 2$p_{3/2}$ and Zn 2$p_{1/2}$ components, respectively (spectrum (i) in Figure 4a). The electronic state of Zn 2$p_{3/2}$ is associated with the Zn^{2+} ions in the ZnO lattice [7,28]. In addition, the Zn 2p XPS spectra of the two impurity-doped samples showed slight shifts toward the lower binding region due to changes in the strength of the metal–oxygen bond energy because of the incorporation of N and Ga–N dopants into the ZnO nanocrystals.

As shown in Figure 4b, the recorded O 1s spectra were obviously broad and asymmetric on the high binding energy side (solid lines). Each O 1s spectrum was deconvoluted to three distinct sub-curves (dashed lines) by Gaussian curve fitting and denoted with O_I, O_{II}, and O_{III} for representing three types of oxygen levels in oxide thin film samples [29,30]. The low binding energy component (O_I) exhibited the strongest intensity, was centered at 529.6 eV, and was associated with O^{2-} ions bonded with metal ions (Zn^{2+} and/or Ga^{3+}), such as Zn–O bonds, in the ZnO lattice. The component at the middle binding energy (O_{II}) was centered at 531.1 eV, usually associated with the loosely bound oxygen, and related to the oxygen vacancy concentration. The component at the high binding energy region (O_{III}), with a weak signal, was ascribed to chemisorbed or dissociated oxygen species, such as hydroxyl groups (–OH) bonded to the surfaces and/or the grain boundaries of the polycrystalline oxide films. The calculated O_{II}/O_{total} intensity ratios for the ZnO, ZnO:N, and ZnO:Ga–N nanostructured thin films were 20.44%, 21.30%, and 27.46%, respectively. Such intensity ratios are suggested for evaluating the relative concentrations of oxygen vacancies between the oxide thin film samples. The calculated results reveal that the co-doped ZnO thin film had a higher O_{II}/O_{total} ratio than those of the un-doped and single-doped ZnO thin films, indicating that weaker metal–oxygen bonds formed in the former sample than in the latter samples. The reduced strength in the metal–oxygen bonds can be attributed to the Ga doping in ZnO:N crystals, which in turn led to a significant increase in the oxygen vacancy concentration in the ZnO:Ga–N thin film.

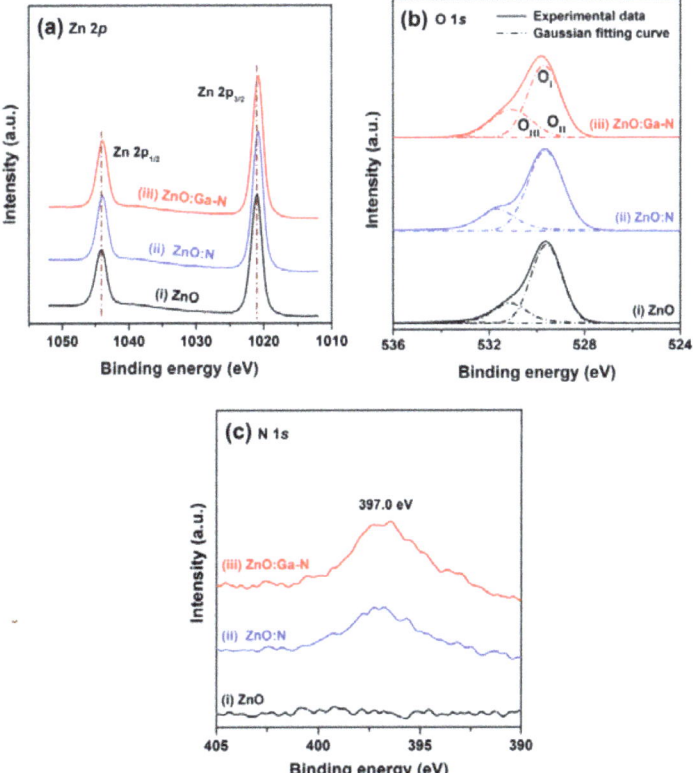

Figure 4. X-ray photoelectron spectroscopy (XPS) spectra of (**a**) Zn 2p, (**b**) O 1s, and (**c**) N 1s of the three ZnO-based thin films.

Figure 4c shows the N 1s XPS core level spectra and the single peak centered at about 397.0 eV, which was related to the formation of the No–Zn bond [23,31] and indicated that N ions were incorporated into the ZnO lattices of the N doped and Ga–N co-doped thin films. It was also demonstrated that Ga–N co-doping enhanced the solubility of the acceptor dopant (N) in ZnO nanocrystals (spectrum (iii) in Figure 4c. Ravichandran et al. suggested that the presence of N in the energy region, about 397.3 eV, is indicative of p-type conductivity [32]. In addition, the Ga 2p spectrum of the Ga–N co-doped sample is presented in Figure S3. The peak spectrum located at 1117.4 eV corresponds to the electrical state of Ga 2$p_{3/2}$ and is ascribed to Ga^{3+} in gallium(III) oxide [33]. The actual contents of N and Ga + N in the ZnO:N and ZnO:Ga–N thin film samples were estimated by XPS analysis to be 0.52 at.%, and 1.1 at.% + 0.76 at.%. The examined results confirm that the N and Ga + N were successfully substituted into the ZnO nanocrystals and the solubility of desirable accept dopant (N) was enhanced through co-doping of Ga and N.

The effects on the optical characteristics of the incorporation N and Ga–N in ZnO thin films were investigated by optical transmission measurement. Figure 5a shows three typical optical transmittance spectra, which were recorded from the samples of glass/oxide thin films in the wavelength range of 200–1000 nm. These transmittance spectra exhibited high transparency (>85.0%) in the visible region and displayed an abrupt increase in absorption in the near-UV spectral range. The character of the optical absorption edge (band-edge absorption) is related to direct optical excitation from the valence band to the conduction band and the magnitude of the bandgap energy. The slight shift toward the short wavelength region of the absorption edge in the two impurity-doped thin film samples may be associated with the finer microstructures.

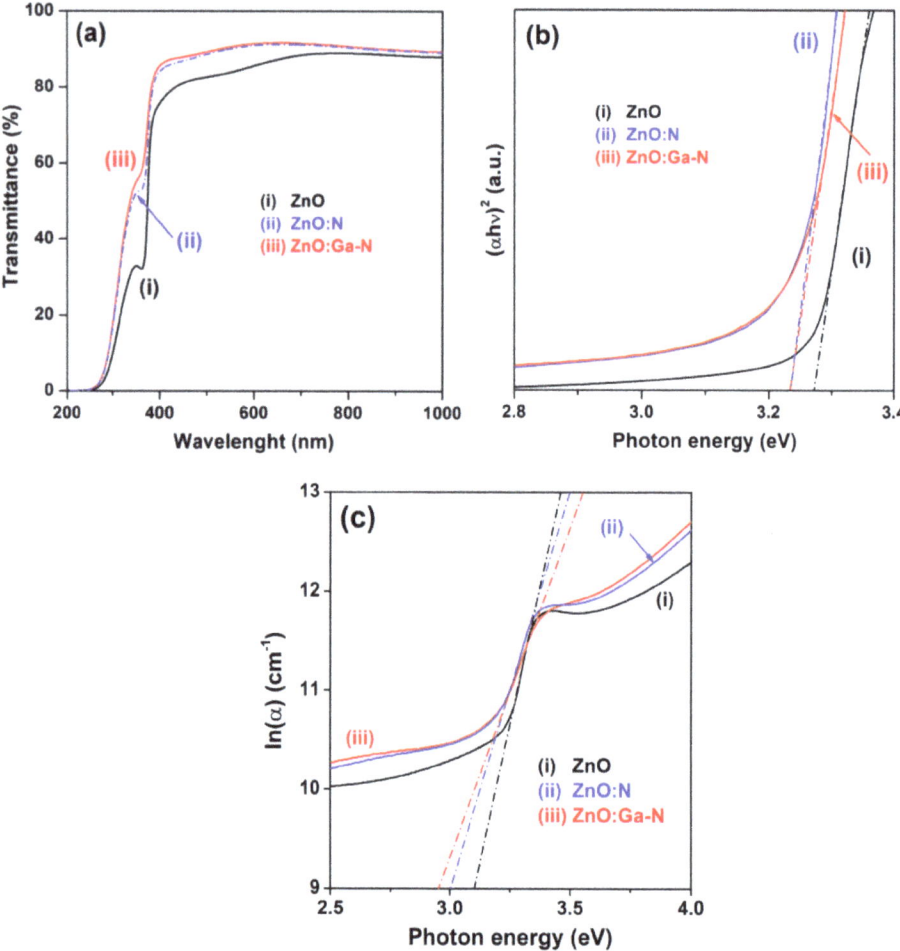

Figure 5. (a) Optical transmission spectra, (b) plot of $(\alpha h\nu)^2$ versus photo energy ($h\nu$), and (c) plot of $\ln(\alpha)$ versus photon energy ($h\nu$) of ZnO, ZnO:N, and ZnO:Ga–N thin film samples.

The recorded optical transmittance spectra indicated that the ZnO:N and ZnO:Ga–N samples exhibited a similar transmission characteristic and had higher transparency than the ZnO sample, covering the visible light region due to the former (impurity-doped samples) being thinner and having a flatter free surface (see Figures 2 and 3) compared with the latter (undoped sample). The average transmittances of undoped and impurity-doped ZnO thin film samples in the visible region (400 to 800 nm) were about 85.0% and approached 90.0% (the fourth column in Table 1). It is known that the optical transmittance of functional oxide thin films is strongly affected by film thickness, surface roughness, densification, crystallinity, and structural homogeneity.

The absorption coefficient (α), one of the intrinsic optical parameters of functional oxide films, can serve as a rough estimation from measured transmittance data with the Beer–Lambert law ($\alpha = [\ln(1/T)]/d_f$, where d_f is the thickness of the examined thin film sample, by assuming reflectance to be negligible ($R = 0$). The calculated results are listed in Table 1. These results reveal that the average absorption coefficient of the undoped sample approached 1.9×10^4 cm^{-1} in the visible region and that an abrupt increase in the absorption coefficient (>26.3%) occurred in the two impurity-doped samples due to the microstructural changes and crystallinity degradation as well as impurity band formation. For optoelectronic materials, the relationship between the absorption coefficient (α) and optical bandgap (E_g) can be expressed as the following relation [34]:

$$\alpha(h\nu) = A(h\nu - E_g)^n \qquad (1)$$

where h is Planck's constant, ν is the frequency of incident photons, the coefficient A is an energy-independent proportionality constant related to the band tailing states, and the exponent n varies according to the energy-band structure of the examined optoelectronic material. For direct allowed transitions, the value of n is 1/2. Figure 5b shows the plot of variation of $(\alpha h\nu)^2$ with the photon energy (eV) (namely, the Tauc plot), which was acquired from the optical transmittance spectra. A Tauc plot is used to determine the optical bandgap energy in semiconductors by extrapolating the straight-line portion near the onset of the absorption edge of the absorption curve to intercept the photon energy axis. The estimated optical bandgap energy of the impurity-doped ZnO thin films was 3.24 eV, which was slightly narrower than that of the undoped ZnO thin film (3.27 eV). Narrowing of the optical bandgap has been reported for ZnO:N, ZnO:Al–N, and ZnO:In–N thin films. That feature can be explained by the change in crystallite size and the formation of shallower impurity (acceptor and donor) states.

The Urbach energy (E_U) width has been considered for the study of the structural disorder and band-tail states of semiconductor thin films with impurity doping [22,35]. It can be estimated by the following expression [36]:

$$\alpha = \alpha_0 \exp(h\nu/E_U), \qquad (2)$$

where α_0 is a constant. We determined the Urbach energy from the plot of ln (α) versus photo energy (hν) (Figure 5c) from the inverse of the slope of the linear segment of the optical absorption spectrum and found that the E_U values increased from 89.2 meV (ZnO) to 123.8 meV (ZnO:N) and 148.5 meV (ZnO:Ga–N). These changes demonstrated that the crystallinity degraded with increases in the dopant species.

Polycrystalline ZnO thin films exhibit strongly stimulated emissions through excitons even at room temperature, which is one of their unique optical properties [37]. Two emission signals appeared in each RT PL spectrum (Figure 6), including a sharp and intense UV emission peak centered at around 379 nm and a weak broad emission band in the visible region. The near-band-edge (NBE) UV emission is created by excitonic emission. A similar UV emission peak (380 nm) has been reported for ZnO thin films deposited on sapphire by sol-gel technique [32]. It is well known that the broad emission in the visible region is due to deep-level emission related to intrinsic defects such as Zn interstitials and O vacancies. It is worth noting that the decrease in the intensity of NBE emission after doping with N and Ga–N indicated the degeneration of the films' crystallinity and was ascribed to the increment of non-radiative recombination due to effects induces by N [22,24]. The N and Ga–N doping also led to a slight shift in the NBE emission peak as noted in the optical transmittance spectra in the near-UV spectral region. Moreover, the suppression of the visible emission may be attributed to the compensation of intrinsic defects by the acceptor dopants (spectra (ii) and (iii) in Figure 6). This feature confirmed the greater solubility and stability of the N dopant incorporated in the co-doped ZnO thin film.

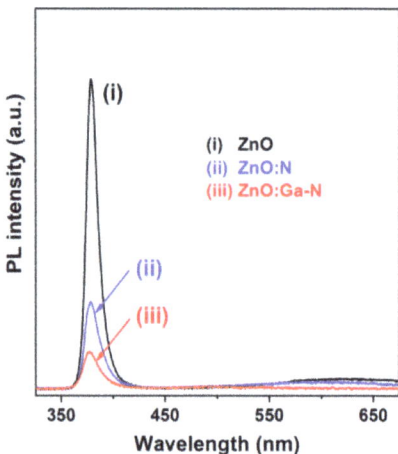

Figure 6. Comparison of the room-temperature photoluminescence (PL) spectra of ZnO, ZnO:N, and ZnO:Ga–N thin film samples.

Various electrical parameters, including the carrier type, carrier concentration, Hall mobility, and electrical resistivity, of the ZnO-based semiconductor thin films, measured by Hall-effect measurement, are presented in Table 2. It was confirmed that the pure ZnO thin films had n-type characteristics with a mean electron concentration of 3.19×10^{15} cm^{-3}, a mean carrier mobility of 20.77 cm^2/V·s, and a mean resistivity of 94.32 Ω·cm. The measurement results also indicate that the as-prepared ZnO:N thin films exhibited a mean hole concentration of 1.83×10^{15} cm^{-3}, a mean hole mobility of 8.85 cm^2/V·s, and a mean resistivity of 385.4 Ω·cm. This change in conductivity type from n to p was related to the incorporation of N dopant into ZnO, which induced shallow acceptors of N_O. However, the hole concentration and resistivity should be improved to achieve device-quality p-type oxide films. It is noted that the mean hole concentration of the p-type ZnO:Ga–N thin films approached 4.0×10^{17} cm^{-3}; this magnitude is comparable to those in previous reports on sol-gel-derived Al–N and In–N co-doped p-type ZnO thin films [22,24]. In addition, the mobility of the co-doped sample was only about 3.0 cm^2/V·s, possibly due to the formation of the Ga–2N complex; this low mobility caused carrier scattering and restricted the hole mobility. Thus, the hole mobility of the p-type ZnO semiconductor thin films for device applications should be improved in future work.

Table 2. Electrical characteristics of the three ZnO-based thin films and the electrical stability of p-type ZnO:N and ZnO:Ga–N thin films.

Sample	Time of Measurement	Carrier Type *	Carrier Concent. (cm^{-3})	Mobility (cm^2/V·s)	Resistivity (Ω·cm)
ZnO	As-deposited	n	-3.19×10^{15}	20.77	94.32
ZnO:N	As-deposited	p	1.83×10^{15}	8.85	385.4
	After 14 days	p	5.21×10^{14}	7.53	1.59×10^3
	After 21 days	n	-2.47×10^{14}	7.87	3.21×10^3
	After 28 days	n	-7.04×10^{13}	7.42	1.19×10^4
ZnO:Ga–N	As-deposited	p	3.96×10^{17}	3.10	5.09
	After 14 days	p	7.52×10^{16}	2.31	36.0
	After 21 days	p	2.28×10^{16}	2.05	133.5
	After 28 days	n	-3.92×10^{15}	4.18	380.9

* The symbols of n and p represent the type of major carrier concentration. (n: electron and p: hole).

Ye et al. summarized the electrical properties of several types of group III (B, Al, Ga, and In)–N co-doped p-type ZnO thin films and found that the hole concentration density, Hall mobility, and electrical resistivity were in the ranges of 3×10^{16} to 5×10^{18} cm^{-3}, 0.1 to 15 cm^2/V·s, and 0.015

to 2×10^2 Ω·cm, respectively [7]. The main considerations of the acceptor-donor co-doping method for achieving p-type conduction in ZnO thin films are to lower the ionization energy of acceptors and donors, enhance the solubility of acceptor dopants [38], and reduce the self-compensation effect to create shallower acceptor levels (falling within the range of 0.1–0.2 eV above the valence-band maximum (VBM)).

For verification of the electrical stability of the p-type oxide thin films, the electrical characteristics were measured in time frames of two to four weeks with periods of 7 days (Table 2). We found that these p-type ZnO:N thin films reverted to n-type over time. The Ga–N co-doped ZnO thin films exhibited more stable p-type conductivity at 3 weeks than did the N-doped ZnO thin films (<three weeks) due to the formation of the Ga_{Zn}–N_O pairs, which could offer more efficient incorporation of N dopants and activation of acceptors, and hence the creation of impurity bands above the VBM of ZnO, which increased the hole concentration. A comparison of the electrical characteristics in previous reports on solution-processed group III element (B, Al, In, and Ga)–N co-doped p-type ZnO thin films with those of the developed ZnO:Ga–N thin films in this work is provided in Table 3. It was found that the electrical properties of ZnO-based thin films fabricated by spray pyrolysis technique are better than those of oxide films prepared by sol-gel process.

Table 3. Comparison of the electrical characteristics of solution-processed group III elements-N co-doped p-type ZnO thin films with the developed ZnO:Ga–N thin film in this study.

Dopant	Method	Hole Concent. (cm^{-3})	Mobility ($cm^2/V \cdot s$)	Resistivity (Ω·cm)	Stability	Ref.
B–N	Spray pyrolysis	5.89×10^{18}	12.86	8.54×10^{-2}	20 days	[18]
Al–N	Spray pyrolysis	2.30×10^{18}	0.94	2.89	-	[28]
Al–N	Sol-gel	2.0×10^{17}	1.6	19.0	-	[24]
Al–N	Sol-gel	8.11×10^{16}	9.97	77.0	-	[14]
In–N	Spray pyrolysis	2.0×10^{19}	3.0	7.70×10^{-2}	30 days	[38]
In–N	Sol-gel	9.81×10^{17}	0.60	1.58×10^{-1}	-	[22]
Ga–N	Spray pyrolysis	3.82×10^{19}	4.34	3.97×10^{-2}	30 days	[19]
Ga–N	Sol-gel	3.96×10^{17}	3.10	5.09	>21 days	This work

The authors examined the optical and electrical properties of the three kinds of sol-gel-derived ZnO-based thin films and connected them with the structural characteristics. Besides comparing the physical properties of the oxide thin films, our experimental investigations clarified the stability of p-type ZnO:N and ZnO:Ga–N thin films. Our study indicates that the p-type ZnO:Ga–N semiconductor thin films exhibited good electrical stability, without significant degeneration of the electrical characteristics, for at least three weeks. We have successfully fabricated Ga–N co-doped ZnO transparent semiconductor thin film, which may be applicable as a stable p-type layer in the development of ZnO-based homojunction or heterojunction electronic and optoelectronic devices.

4. Conclusions

Electrically stable p-type ZnO transparent semiconductor thin films have been deposited on alkali-free glass substrates with Ga and N co-doping by the sol-gel spin-coating process. XRD patterns showed that these sol-gel-derived ZnO-based thin films were polycrystalline with a single hexagonal structure, and the microstructural features were significantly changed by the impurity doping. XPS analysis confirmed the incorporation of N and Ga–N into the two kinds of impurity-doped thin films and demonstrated that Ga and N co-doping enhanced the solubility of N dopant in ZnO thin films. The N-doped and Ga–N co-doped ZnO thin films both exhibited high transparency (~90%) in the visible spectrum. It was confirmed that the nature of conductivity of ZnO semiconductor thin films changed from n-type to p-type when the N dopant was incorporated in the ZnO nanocrystals. In the present study, we succeeded in realizing good conductivity in p-type ZnO thin film through simultaneous co-doping of Ga and N, achieving a hole concentration of 3.96×10^{17} cm^{-3}, mobility of 3.1 $cm^2/V \cdot s$, and resistivity of 5.09 Ω·cm. The p-ZnO:Ga–N transparent semiconductor thin films exhibited electrical stability for almost four weeks.

Supplementary Materials: The following are available online at http://www.mdpi.com/2079-6412/10/11/1069/s1, Figure S1: Plane-view FE-SEM micrographs of ZnO-based thin films: (a) undoped, (b) N-doped, and (c) Ga–N co-doped samples, Figure S2: 2D SPM images of ZnO-based thin film surfaces: (a) undoped, (b) N-doped, and (c) Ga–N co-doped samples, Figure S3: Core level XPS spectrum of Ga 2p of the Ga–N co-doped ZnO thin film.

Author Contributions: Conceptualization, C.-Y.T. and W.-Y.C.; methodology, C.-Y.T. and W.-Y.C.; investigation and resources, C.-Y.T.; data curation, C.-Y.T. and W.-Y.C.; writing—original draft preparation and writing—review and editing, C.-Y.T.; project administration, C.-Y.T. All authors have read and agreed to the published version of the manuscript.

Funding: This research received no external funding.

Acknowledgments: The authors gratefully acknowledge the Precision Instrument Support Center of Feng Chia University for providing XRD, FE-SEM, and SPM characterization facilities and the Instrument Center of National Chung Hsing University, Taichung, Taiwan, for help with the XPS measurement.

Conflicts of Interest: The authors declare that there are no known conflict of interest associated with this publication and there is no significant financial support for this study that could have influenced its outcome.

References

1. Özgür, Ü.; Hofstetter, D.; Morkoç, H. ZnO devices and applications: A review of current status and future prospects. *Proc. IEEE* **2010**, *98*, 1255–1268. [CrossRef]
2. Özgür, Ü.; Alivov, Y.I.; Liu, C.; Teke, A.; Reshchikov, M.A.; Doğan, S.; Avrutin, V.; Cho, S.J.; Morkoç, A.H. A comprehensive review of ZnO materials and devices. *J. Appl. Phys.* **2005**, *98*, 041301. [CrossRef]
3. Wang, Z.; Nayak, P.K.; Caraveo-Frescas, J.A.; Alshareef, H.N. Recent developments in p-type oxide semiconductor materials and devices. *Adv. Mater.* **2016**, *28*, 3831–3892. [CrossRef] [PubMed]
4. Yadav, H.M.; Otari, S.V.; Bohara, R.A.; Mali, S.S.; Pawar, S.H.; Delekar, S.D. Synthesis and visible light photocatalytic antibacterial activity of nickel-doped TiO_2 nanoparticles against Gram-positive and Gram-negative bacteria. *J. Photochem. Photobiol. A Chem.* **2014**, *294*, 130–136. [CrossRef]
5. Shang, Z.-W.; Hsu, H.-H.; Zheng, Z.-W.; Cheng, C.-H. Progress and challenges in p-type oxide-based thin film transistors. *Nanotechnol. Rev.* **2019**, *8*, 422–443. [CrossRef]
6. Zhang, S.B.; Wei, S.-H.; Zunger, A. Intrinsic n-type versus p-type doping asymmetry and the defect physics of ZnO. *Phys. Rev. B* **2001**, *63*, 075205. [CrossRef]
7. Ye, Z.; He, H.; Jiang, L. Co-doping: An effective strategy for achieving stable p-type ZnO thin films. *Nano Energy* **2018**, *52*, 527–540. [CrossRef]
8. Yan, Y.; Li, J.; Wei, S.-H.; Al-Jassim, M.M. Possible approach to overcome the doping asymmetry in wideband gap semiconductors. *Phys. Rev. Lett.* **2007**, *98*, 135506. [CrossRef]
9. Janotti, A.; Van De Walle, C.G. Fundamentals of zinc oxide as a semiconductor. *Rep. Prog. Phys.* **2009**, *72*, 126501. [CrossRef]
10. Tsay, C.-Y.; Lee, W.-C. Effect of dopants on the structural, optical and electrical properties of sol–gel derived ZnO semiconductor thin films. *Curr. Appl. Phys.* **2013**, *13*, 60–65. [CrossRef]
11. Duan, X.M.; Stampfl, C.; Bilek, M.M.M.; McKenzie, D.R. Codoping of aluminum and gallium with nitrogen in ZnO: A comparative first-principles investigation. *Phys. Rev. B* **2009**, *79*, 235208. [CrossRef]
12. Look, D.C.; Claflin, B.; Alivov, Y.I.; Park, S.J. The future of ZnO light emitters. *Phys. Status Solidi* **2004**, *201*, 2203–2212. [CrossRef]
13. Park, C.H.; Zhang, S.B.; Wei, S.-H. Origin of p-type doping difficulty in ZnO: The impurity perspective. *Phys. Rev. B* **2002**, *66*, 073202. [CrossRef]
14. Saravanakumar, B.; Mohan, R.; Thiyagarajan, K.; Kim, S.-J. Investigation of UV photoresponse property of Al, N co-doped ZnO film. *J. Alloys Compd.* **2013**, *580*, 538–543. [CrossRef]
15. Yamamoto, T.; Katayama-Yoshida, H. Solution using a codoping method to unipolarity for the fabrication of p-type ZnO. *Jpn. J. Appl. Phys.* **1999**, *38*, L166–L169. [CrossRef]
16. Joseph, M.; Tabata, H.; Kawai, T. p-type electrical conduction in ZnO thin films by Ga and N codoping. *Jpn. J. Appl. Phys.* **1999**, *38*, L1205–L1207. [CrossRef]
17. Li, W.; Kong, C.; Qin, G.; Ruan, H.; Fang, L. p-type conductivity and stability of Ag–N codoped ZnO thin films. *J. Alloys Compd.* **2014**, *609*, 173–177. [CrossRef]
18. Narayanan, N.; Deepak, N. B–N codoped p type ZnO thin films for optoelectronic applications. *Mater. Res.* **2018**, *21*, 20170618. [CrossRef]

19. Narayanan, N.; Deepak, N. Melioration of optical and electrical performance of Ga–N codoped ZnO thin films. *Z. Für Nat. A* **2018**, *73*, 547–553. [CrossRef]
20. Lu, J.G.; Ye, Z.Z.; Zhuge, F.; Zeng, Y.-J.; Zhao, B.H.; Zhu, L.P. p-type conduction in N–Al co-doped ZnO thin films. *Appl. Phys. Lett.* **2004**, *85*, 3134. [CrossRef]
21. Matsui, H.; Saeki, H.; Tabata, H.; Kawai, T. Role of Ga for Co-doping of Ga with N in ZnO films. *Jpn. J. Appl. Phys.* **2003**, *42*, 5494–5499. [CrossRef]
22. Cao, Y.; Miao, L.; Tanemura, S.; Tanemura, M.; Kuno, Y.; Hayashi, Y. Low resistivity p-ZnO films fabricated by sol-gel spin coating. *Appl. Phys. Lett.* **2006**, *88*, 251116. [CrossRef]
23. Bian, J.M.; Li, X.M.; Gao, X.D.; Yu, W.D.; Chen, L.D. Deposition and electrical properties of N–In codoped p-type ZnO films by ultrasonic spray pyrolysis. *Appl. Phys. Lett.* **2004**, *84*, 541–543. [CrossRef]
24. Dutta, M.; Ghosh, T.; Basak, D. N doping and Al–N co-doping in sol-gel ZnO films: Studies of their structural, electrical, optical, and photoconductive properties. *J. Electron. Mater.* **2009**, *38*, 2335–2342. [CrossRef]
25. Janotti, A.; Van De Walle, C.G. Oxygen vacancies in ZnO. *Appl. Phys. Lett.* **2005**, *87*, 122102. [CrossRef]
26. Tsay, C.-Y.; Hsu, W.-T. Comparative studies on ultraviolet-light-derived photoresponse properties of ZnO, AZO, and GZO transparent semiconductor thin films. *Materials* **2017**, *10*, 1379. [CrossRef]
27. Tsay, C.-Y.; Fan, K.-S.; Lei, C.-M. Synthesis and characterization of sol-gel derived gallium-doped zinc oxide thin films. *J. Alloys Compd.* **2012**, *512*, 216–222. [CrossRef]
28. Zhang, X.; Fan, H.; Sun, J.; Zhao, Y. Structural and electrical properties of p-type ZnO films prepared by ultrasonic spray pyrolysis. *Thin Solid Films* **2007**, *515*, 8789–8792. [CrossRef]
29. Park, G.C.; Hwang, S.M.; Lee, S.M.; Choi, J.H.; Song, K.M.; Kim, H.Y.; Eum, S.-J.; Jung, S.-B.; Lim, J.H.; Joo, J. Hydrothermally grown In-doped ZnO nanorods on p-GaN films for color-tunable heterojunction light-emitting-diodes. *Sci. Rep.* **2015**, *5*, 10410. [CrossRef]
30. Futsuhara, M.; Yoshioka, K.; Takai, O. Structural, electrical and optical properties of zinc nitride thin films prepared by reactive rf magnetron sputtering. *Thin Solid Films* **1998**, *322*, 274–281. [CrossRef]
31. Zhang, J.P.; Zhang, L.D.; Zhu, L.Q.; Zhang, Y.; Liu, M.; Wang, X.J.; He, G. Characterization of ZnO:N films prepared by annealing sputtered zinc oxynitride films at different temperatures. *J. Appl. Phys.* **2007**, *102*, 114903. [CrossRef]
32. Ravichandran, C.; Srinivasan, G.; Lennon, C.; Sivanathan, S.; Kumar, J. Investigations on the structural and optical properties of Li, N and (Li, N) co-doped ZnO thin films prepared by sol-gel technique. *Mater. Sci. Semicond. Process.* **2010**, *13*, 46–50. [CrossRef]
33. Rakhshani, A.E.; Bumajdad, A.; Kokaj, J.; Thomas, S. Structure, composition and optical properties of ZnO:Ga films electrodeposited on flexible substrates. *Appl. Phys. A* **2009**, *97*, 759–764. [CrossRef]
34. Huang, K.; Tang, Z.; Zhang, L.; Yu, J.; Lv, J.; Liu, X.; Liu, F. Preparation and characterization of Mg-doped ZnO thin films by sol-gel method. *Appl. Surf. Sci.* **2012**, *258*, 3710–3713. [CrossRef]
35. Melsheimer, J.; Ziegler, D. Band gap energy and Urbach tail studies of amorphous, partially crystalline and polycrystalline tin dioxide. *Thin Solid Films* **1985**, *129*, 35–47. [CrossRef]
36. Mia, M.N.H.; Pervez, M.F.; Hossain, M.K.; Rahman, M.R.; Uddin, M.J.; Al Mashud, M.A.; Ghosh, H.K.; Hoq, M. Influence of Mg content on tailoring optical bandgap of Mg-doped ZnO thin film prepared by sol-gel method. *Results Phys.* **2017**, *7*, 2683–2691. [CrossRef]
37. Ryu, Y.R.; Zhu, S.; Look, D.C.; Wrobel, J.M.; Jeong, H.M.; White, H.W. Synthesis of p-type ZnO films. *J. Cryst. Growth* **2000**, *216*, 330–334. [CrossRef]
38. Yu, C.-C.; Lan, W.-H.; Huang, K.-F. Indium-nitrogen codoped zinc oxide thin film deposited by ultrasonic spray pyrolysis on n-(111) Si substrate: The effect of film thickness. *J. Nanomater.* **2014**, *2014*, 1–7. [CrossRef]

Publisher's Note: MDPI stays neutral with regard to jurisdictional claims in published maps and institutional affiliations.

© 2020 by the authors. Licensee MDPI, Basel, Switzerland. This article is an open access article distributed under the terms and conditions of the Creative Commons Attribution (CC BY) license (http://creativecommons.org/licenses/by/4.0/).

Article

Influence of the Growth Ambience on the Localized Phase Separation and Electrical Conductivity in SrRuO$_3$ Oxide Films

Hsin-Ming Cheng

Organic Electronics Research Center and Department of Electronic Engineering, Ming Chi University of Technology, New Taipei City 24301, Taiwan; SMCheng@mail.mcut.edu.tw

Received: 18 August 2019; Accepted: 16 September 2019; Published: 18 September 2019

Abstract: Perovskite SrRuO$_3$ (SRO) epitaxial thin films grown on SrTiO$_3$ (STO) (001) have been synthesized using pulsed laser deposition (PLD) under a series of oxygen pressures. High quality and conductive SRO thin films on STO have been achieved at 10^{-1} Torr oxygen pressure with the epitaxial relation of (110)<001>$_{SrRuO_3}$//(001)<010>$_{SrTiO_3}$. The lattice parameters of the thin films exhibit huge expansion by reducing the ambience (~10^{-7} Torr) during deposition, and the resistance increases by about two orders higher as compared with the low oxide pressure ones. The rise of resistivity can be ascribed to not only the deficiency of Ru elements but also the phase transformation inside SRO thin films. The correlation of growth ambience on the structural transition and corresponding resistivity of epitaxial oxide thin films have been explicitly investigated.

Keywords: pulsed laser deposition; functional oxide; phase transformation; electrical conductivity

1. Introduction

Complex ruthenium oxide materials are fascinating because they possess a range of physical properties including being superconductive [1,2] and ferromagnetic [3–11], and they also have metallic properties [12–15]. Among the ruthenium oxide family, metallic oxide SrRuO$_3$ (SRO) has attracted much attention because of its functionality as a conducting electrode in the integration of perovskite oxide devices such as magnetic tunnel junctions [16,17]. SRO, a 4d transition metal oxide with a Curie temperature of 161 K, is a Pbnm orthorhombic structure with lattice parameters a = 5.57 Å, b = 5.53 Å, and c = 7.86 Å [18]. Due to a small distortion with respect to the ideal cubic structure, SRO is also regarded as pseudo-cubic with a parameter of 3.93 Å [19,20]. In addition, the conductivity of SRO comes from a π^* narrow-type conduction band, which originates from a strong hybridization of oxygen 2p-derived states with ruthenium d states [21,22]. The resistivity of SRO thin film is about 200 μΩ-cm at room temperature (RT) [15]. Structural compatibility with perovskite oxides and metallic conductivity make SRO widely used for electrodes in oxide-based devices.

As an itinerant ferromagnetic oxide, in the past twenty-five years, SRO thin films have been proposed that can be successfully deposited on vicinal substrates such as SrTiO$_3$ (STO) through heteroepitaxial step-flow growth [23–26]. This model has offered a rational explanation for the morphological phase diagram, step bunching, and island formation of SRO surfaces. However, in order to obtain high-quality functional oxide films, the control of the atmosphere, especially the oxygen ambience, plays an essential role in the aspects of defects [6,27], stoichiometry [14], oxidations [28], and reductions [28]. Furthermore, oxygen partial pressure dramatically causes phase transformations as well [7,11–13].

For SRO functional oxide materials, the ambience effects on their surface stability have been studied by post-annealing at temperatures higher than 600 °C under different ambiences [28,29]. Lee et al. reported that SRO will decompose below an oxygen pressure of 10^{-4} Torr at 720 °C [29].

Shin et al. further listed the decomposition reactions of SRO at a range of oxygen pressures and temperatures based on thermodynamic theory [28]. Although these reactions are complicated due to the fact that nonstoichiometric SrO will not coexist in equilibrium with SRO, the report also pointed out that ruthenium seems to lose more than strontium at high vacuum (10^{-7} Torr) and high temperature (700 °C). Siemons et al. reported that either too low or too high oxygen activity could lead to the deficiency of Ru in SRO thin films, which increases the resistivity [14]. On the other hand, Lee et al. recently reported a close correlation between the phase transitions and oxygen evolution reaction of SRO epitaxial thin films by systematically introducing Ru–O vacancies [30]. However, even though previous reports have revealed the resistivity of SRO thin films as entire thin film, namely the macro-scale measurement, there are few investigations that mention the microstructural and local electrical properties of SRO with various growth ambience. As a result, to understand the complexity between the process and the conductivity that can be used for practical application in the future, the subject related to the growth ambience of SRO thin films needs to be diligently discussed. In this study, we control the oxygen pressure from 10^{-1} to 10^{-7} Torr in order to systematically study the influence of growth ambience on the structure and conductivity of SRO thin films. The crystallization and the corresponding conductivity of SRO thin films vary significantly as a function of the ambient environments. Furthermore, the variations of crystallization from the deficiency of Ru elements, surface morphology, degree of stoichiometry, and the microstructure caused by phase transformation in SRO thin films are also investigated.

2. Experiment

Epitaxial thin films of SRO were deposited on $SrTiO_3$ (001) substrate using pulsed laser deposition (PLD, DCA PLD500, Turku, Finland) with a target to substrate distance of ~55 mm using a KrF excimer laser (λ = 248 nm, pulse duration 4 ns). The samples were grown in a vacuum chamber with a background pressure of 10^{-8} Torr at 700 °C using pulsed energy of 90 mJ with a repetition rate of 8 Hz. The fluence was refined with an attenuator to get 6 J/cm^2 on the target, with a laser spot size of 1.5 mm^2. The thickness of the SRO thin films was controlled to around 300 nm. After growth, the samples were placed in the original position and remained there to cool naturally. The ambient oxygen during growth was varied from 10^{-1} to 10^{-7} Torr, while other conditions were kept the same in order to confirm how oxygen pressure affected the structural and electrical properties. The crystal structure of the grown films was identified using a high-resolution four-circle X-ray diffractometer (Bede D1, HRXRD, Bede, Durham, UK) with Cu-Kα_1 radiation. The surface morphology of the samples was characterized by field emission scanning electron microscopy (FESEM, JEOL-6500, Tokyo, Japan), and their ingredients were analyzed using an energy dispersive spectrometer (EDS) operated at 5 kV. Cross sectional TEM specimens were prepared by focused ion beam (FIB) and characterized by field emission transmission electron microscope (FETEM, JEOL JEM-2100F) operated at 200 kV. The electrical resistivity at RT was measured using a standard four-probe technique. Atomic force microscopy (AFM, Bruker Innova, SPM, Billerica, MA, USA), working in contact mode, was used to characterize the surfaces of the films. Local electrical conductivity was measured using high-resolution conductive AFM (C-AFM) by applying a bias voltage of 0.2 V between the C-AFM tip and the surrounding silver colloid.

3. Results and Discussion

X-ray reciprocal space mapping (RSM) along the θ–2θ scan and the rocking curve θ around STO (002) and SRO (220) reflections were used to analyze the structure of SRO on STO (001). Figure 1 shows the RSM results for SRO thin films grown under an oxygen pressure of 10^{-1} Torr (Sample A) and 10^{-6} Torr (Sample C), respectively. Only SRO (220) and STO (002) reflections are observed, revealing that the SRO (110) thin films are parallel to the STO (001). These results are consistent with previous reports. As can be seen in Figure 1, Sample A is different from Sample C in both the peak position and the full width at half maximum (FWHM) of the θ-rocking curve along the surface normal direction. The peak position and FWHM of Sample A are 46.169° and 0.074°, respectively. These results indicate that SRO thin films have a d_{110}-spacing of 3.928 Å, for which the value is very close to its bulk value (3.925 Å). The sharp rocking curve also reveals an excellent crystallization of the SRO thin films.

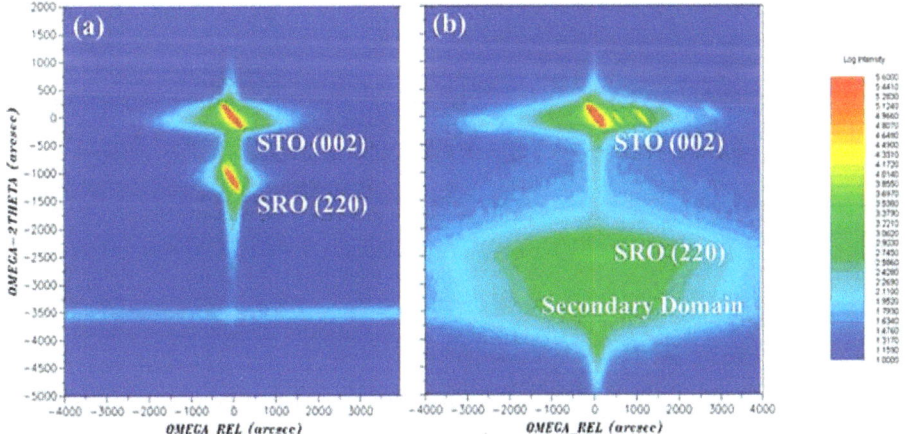

Figure 1. RSM around the normal surface of (002) reflection for SrTiO$_3$ (STO) and (220) reflection for SrRuO$_3$ (SRO) thin films of (**a**) Sample A and (**b**) Sample C, grown at 10^{-1} and 10^{-6} Torr, respectively. (The unit of arcsec is equal to 1 over 3600 degree).

Figure 2 illustrates the φ-scans across the off-normal SRO (112) and STO (011) reflections of Sample A. It reveals a sharp width of 0.0471° (0.0135° for STO substrate) with a near four-fold symmetry for SRO (112). It indicates a very high quality of SRO thin film by using the high oxygen pressure of 10^{-1} Torr. Furthermore, the results point out that the in-plane epitaxial relationship follows $\{001\}_{SrRuO_3} \| \{010\}_{SrTiO_3}$, which is consistent with the work of Kan et al. [14].

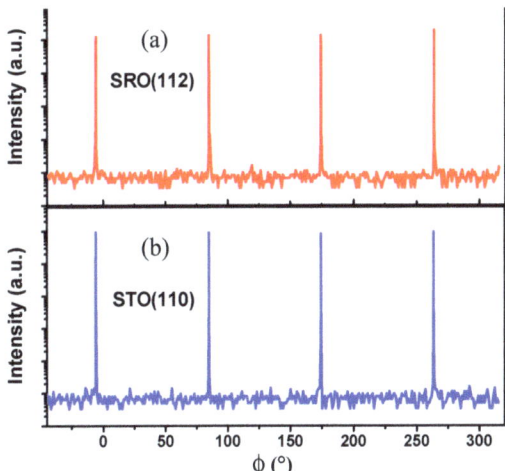

Figure 2. The φ-scans across the off-normal SrRuO$_3$ reflection (**a**) and SrTiO$_3$ reflection (**b**) of Sample A.

On the contrary, X-ray diffraction also shows that the lattice parameter and FWHM expand a lot in Sample **C**, which was grown under a lower oxygen pressure. The calculated lattice parameter and rocking curve of Sample **C** are 3.959 Å and 0.2725°, respectively. Consequently, the results indicate that the crystal structure of SRO thin films is very sensitive to oxygen pressure during growth. In this work, it was found that a high oxygen pressure of about 100 mTorr facilitated the growth of high crystal-quality SRO thin films. At the lower side of Figure 1b, a small elliptical region is found in Sample C, which indicates that SRO is composed of two domains with different lattice constants (3.959 and 3.987 Å). Sample D also shows the multi-domain phenomenon, while the lattice constants of sample D are larger than those of Sample C, as shown in Figure S1. These secondary domains might be formed due to the instability of SRO at high temperature and simultaneously high vacuum, and they would be almost undetectable while the oxygen pressure is above 10^{-4} Torr.

The detailed variations of lattice constants, strains, and sheet resistances under different growth oxygen pressures are listed in Table 1. The out-of-plane lattice constant expands as the oxygen pressure decreases, and the secondary domains with larger lattice parameters are clearly observed bellow a growth ambience of 10^{-5} Torr. Furthermore, the electrical property changes with the growth ambience. From the results of the four-probe method, the resistivity of Sample B is slightly high but still of the same order as Sample A. However, the resistivity of Samples C and D substantially increase with decreasing oxygen pressure during growth. Sample D, grown without the addition of oxygen, reveals a resistivity about two orders higher than that of Sample A. Therefore, structure and electricity seem to vary with growth ambience in two stages. The lattice constant of SRO thin films increases, but the conductivity decreases gradually while controlling the oxygen pressure from 10^{-1} to 10^{-4} Torr. Several secondary domains are formed, and the conductivity of SRO thin films drops while the oxygen pressure is below 10^{-4} Torr. Therefore, there should be a mechanism responsible for the different variations in the two stages. The mechanism of differences in the decrease rate is discussed in detail below.

Table 1. The lattice parameters, strains, and resistivity of SRO thin films fabricated under different oxygen pressures.

Sample	A	B	C	D
Oxygen pressure (Torr)	10^{-1}	10^{-4}	10^{-6}	10^{-7}
d-spacing of (110) (Å)	3.928	3.947	3.959	3.963
Secondary domain d (Å)	–	–	3.987	4.076
FWHM (Degree)	0.0741	0.2725	0.3190	0.4045
Resistivity (μΩ-cm)	224	600	4380	61,440

3.1. SEM

The morphologies of SRO thin films were measured by SEM, as shown in Figure 3. As can be seen, SRO thin films have some interesting morphologies depending on their growth conditions. Sample A reveals high quality with a relatively flat surface without any obvious particles This means that no secondary domains were formed, which is consistent with the RSM results. However, the surfaces become rough and harsh when the oxygen pressure decreases. Some tiny protrusions have begun to form on the surface of the film while the atmosphere decreases to 10^{-4} Torr, as shown in Figure 3b. There are some uniform muffin-like particles spreading on the flat surfaces of Samples C and D, as shown in Figure 3c,d. The sizes of the muffins strongly depend on the growth pressure. The average diameter and height of Sample C are 0.35 μm and 30 nm, respectively, while those of Sample D are 0.65 μm and 50 nm. This indicates that the formation of the secondary domains as muffin-like regions while the oxygen pressure is below 10^{-4} Torr is consistent with the XRD measurements.

SEM-EDS, with the same electron energy, was used to analyze the composition and stoichiometry of the SRO samples. They both contain Sr, O, and Ru (the Ti signal is coming from the substrate). The calculated ratios of Sr/Ru for different samples are quantized and listed. The results of the analysis show that all films exhibited Ru deficiency throughout the working pressure range. As a result, the composition of SRO thin films cannot be directly controlled by the partial pressure parameter. This is consistent with previous findings that growth in an oxygen atmosphere is due to Ru deficiency [6,7,14].

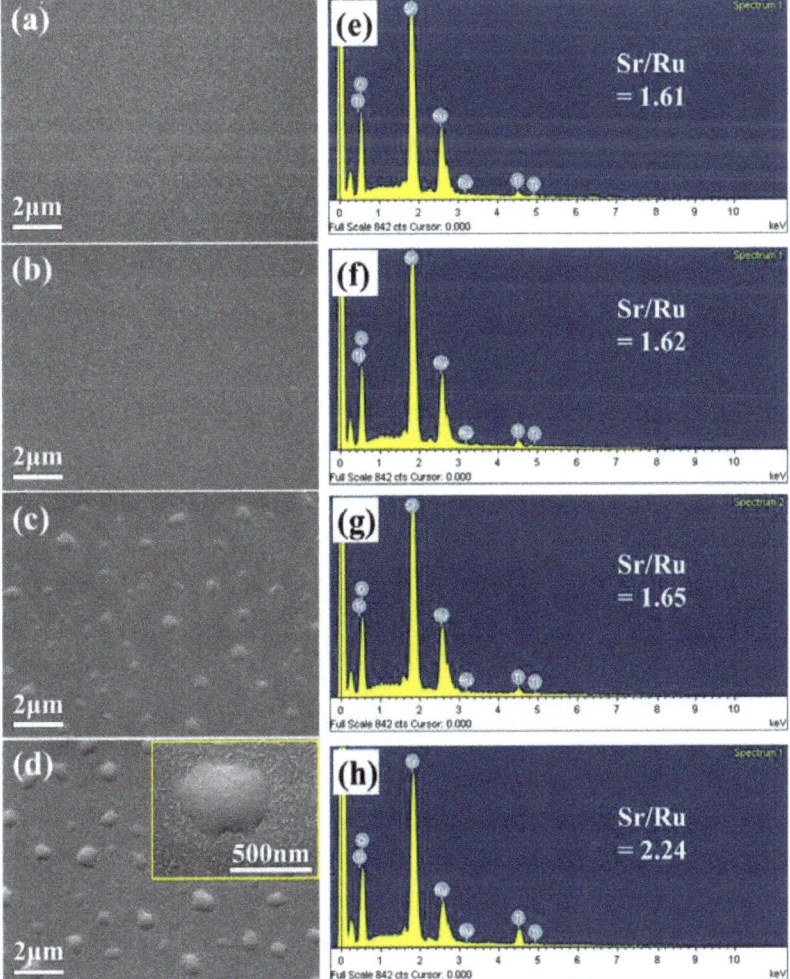

Figure 3. (a–d) are SEM images of Sample A, Sample C, and Sample D, respectively; (e–h) are energy dispersive spectrometer (EDS) images of Sample A, Sample C, and Sample D, respectively. A magnetized image of the muffin-like region is shown in the inset of (d). The corresponding atomic ratios of Sr/Ru are also shown on the right side.

Sample D has the highest ratio of 2.24, indicating that more Ru elements run out by reducing the growth ambience. The change in Sr/Ru could be due to the lattice parameter expansion and the conductivity decrease. Keeble et al. showed that *A*-site or *B*-site vacancy in STO perovskite oxide thin films would lead to *c*-axis lattice parameter expansion, which is consistent with our results [31]. The poor electricity can be ascribed to Ru deficiency, which reduces the conduction band hybridization due to there being less overlap between Ru and O orbitals. X-ray photoemission spectroscopy (XPS) was also carried out to analyze the SRO thin films, as shown in Figure S2. The peaks at the binding energies 464.3, 531.7, and 529.2 eV correspond to the Ru $3p_{3/2}$ defect-like oxygen vacancy and SRO (O 1s) orbitals, respectively. Technically, it is difficult to determine the exact stoichiometry of oxide thin films, but the trend of an increase of oxygen vacancies with decreasing oxygen partial pressure can still be obtained, with the result being similar to recent reports [30].

3.2. TEM

Sample D was studied using TEM because it showed differences in both structural and electrical properties. As shown in Figure 4a, the cross-section image reveals a 300 nm thickness. The black layer at the top of the thin film is Pt (which is used for SEM measurement), and the bottom corresponds to the STO substrate. The SRO thin film clearly segregates into two different regions, which are, respectively, the muffin-like region with a white color (Region 1) and the thin film region with a gray color (Region 2). The structures of Regions 1 and 2 are analyzed in Figure 4b,c, respectively.

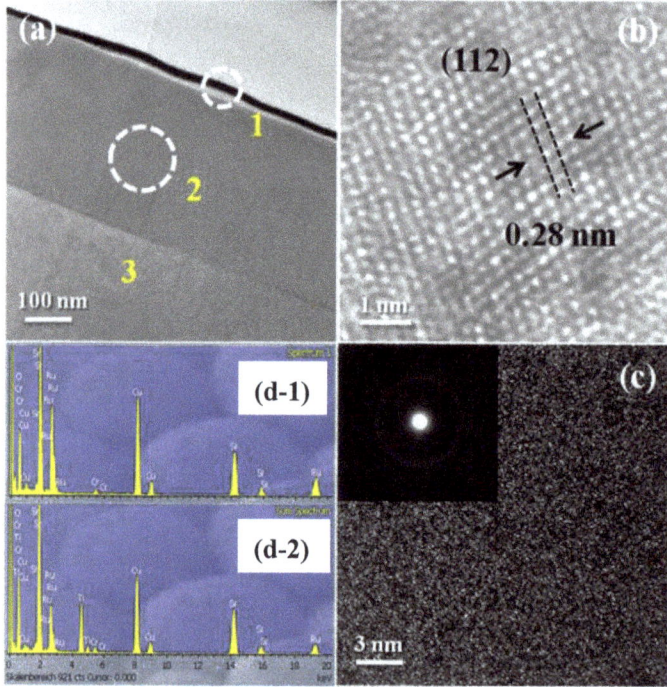

Figure 4. (a) Cross-sectional high-resolution transmission electron microscope (HRTEM) image of Sample D where Regions 1, 2, and 3 are the muffin region, the thin film region, and the substrate, respectively. (b) The magnified TEM of Region 1. (c) The magnified TEM of Region 2 and its corresponding selected area diffraction pattern (SAED). (d) The selected area TEM-EDS of Regions 1 (d-1) and 2 (d-2).

The spacing of the (112) planes in the muffin-like region, shown in Figure 4b, is 2.82 Å. It shows little expansion due to the lack of oxygen during growth. However, the diffraction pattern of Region 2 reveals a polycrystalline structure with random orientation. The lattice constant in Region 2 is 2.94 Å. This indicates that Region 1 is closer to the stoichiometry of $SrRuO_3$ compared with Region 2. Figure 4d shows the TEM-EDS of Regions 1 (the upper image) and 2 (the lower image). The signals of Cu and Cr come from the hole-grids, while that of Ti comes from the substrate. Comparing the upper and lower images, the two regions contain apparently different ratios of ingredients, especially the Ru elements. Region 2 contains less Ru than Region 1. The calculated ratio of Ru/Sr is close to one for the muffin-like regions, whereas it is 0.516 for the thin films.

By combining the ingredients with the diffraction patterns in different areas, the muffin-like regions are regarded as SRO. The thin films, however, should be considered to be $Sr_4Ru_2O_9$ [21] for the following reasons: First, the diffraction patterns of thin film around 2.94 Å is close to the (300) peak of $Sr_4Ru_2O_9$, which is the highest diffraction peak in $Sr_4Ru_2O_9$. Second, the ratio of Ru/Sr is near 0.5. Moreover, the $Sr_4Ru_2O_9$ is less conductive than SRO at room temperature [32]. Shin et al. reported that under a strong reduction condition at 700 °C under a high vacuum of 10^{-7} Torr, some intermediated SrO-rich phases could form and coexist in the SRO thin film [28]. As a result, the formation of $Sr_4Ru_2O_9$ is not beyond expectation.

The degrees of Ru-deficient SRO thin films are well controlled by using a range of oxygen pressures during deposition. The structure transforms from a perfect epitaxial SRO thin film to a random crystallized SrO-rich phase below 10^{-6} Torr. In such a severe condition, SRO splits into the muffin-like regions and the flat regions. The flat regions contain some $Sr_4Ru_2O_9$ components while the muffin-like regions are $SrRuO_3$. SrO-rich phase dominates the structure when both the temperature is high (700 °C) and the oxygen ambience is insufficient (10^{-7} Torr). The formation of SrO-rich phase should influence not only the crystallizations but also the electrical properties of SRO. To analyze the area-dependent-electric properties, we used the scanning probe microscopy technique, which can detect conductive properties in nano-scale size.

3.3. AFM and C-AFM

A conductive AFM (C-AFM) was used as a local probe to analyze the electrical properties of the muffin-like regions and the flat regions in Samples C and D. The schematic diagram of the experimental set-up is shown in Figure 5a. Figure 5b shows the surface morphology of Sample C. The average diameter and height of the muffin-like regions in Sample **C** are 0.35 μm and 30 nm, which are consistent with the SEM results.

The current-mapping images of Samples C and D are shown in Figure 5c,d, respectively. The images apparently reveal brighter colors for flat regions and darker ones for muffin-like regions, indicating a different quantity of current flowing through the two regions. For Sample C, the current in the flat region is about 100 nA, while that in the muffin-like region is less than 50 pA. There is an obvious discontinuity between the muffin-like regions and the nearby flat areas since the currents are prevented from transmitting through their interfaces. This might be due to the electron scattering and the barriers at the interfaces. The interfaces involving a structural transform from SRO to $Sr_4Ru_2O_9$ could contain some defects. Sample D shows the same contrast colors between the muffin-like regions and the flat regions, but the current in the flat regions reduces to 5 nA. This might be a result of the lower stoichiometry in Sample D compared with Sample C.

In Figure 5e, the local I-V curves of Points 1 and 2 represent the muffin-like region and the flat region for Sample C, respectively, and Points 4 and 6 are those for Sample D. The current of Point 2 increases linearly with the applied voltage. A current of up to 100 nA is observed when applying a voltage less than 0.1 V. Point 6 shows a current of 70 nA as the applied voltage increases to 1 V. It again indicates that there is a decrease in conductivity due to the phase transform from SRO to $Sr_4Ru_2O_9$. As shown in the inset of Figure 5e, Points 1 and 4 show currents lower than 50 pA, even when the voltages are over 1 V. The muffin-like regions embedded in $Sr_4Ru_2O_9$ thin films behave as complete isolators without any leakage currents. In contrast, Sample A has a micro-crystalline area on the surface but is relatively flat compared with Samples C and D, and the current-mapping image on its surface performs almost uniformly except for a few areas, as shown in Figure S3. Not following the step-flow model, the samples in this work are relatively rough, indicating a three-dimensional island growth mode characteristic. The coalescence of grain boundaries and step edges begins to dominate at higher pressures, with the results being similar to the previous reports [7,8], but a better conductive SRO thin film with a stochiometric structure can be conducted in a relatively higher oxygen pressure around 10^{-1} Torr. By comparing the C-AFM with the electrical properties listed in Table 1, we can conclude that both the structure and electricity of SRO are strongly affected by the growth ambience. When the oxygen pressure varies from 10^{-1} to 10^{-5} Torr, the conductivity decreases slowly due to a lack of Ru elements. In addition, the out-of-plane lattice parameter of SRO expands as a function of reduced oxygen. In this ambience range, the structure seems more sensitive than the electricity. In the second range, when the oxygen pressure is below 10^{-6} Torr, the resistivity of SRO increases substantially. The process accompanies a phase transition from SRO to $Sr_4Ru_2O_9$. The secondary phase has an intrinsic resistivity higher than that of intrinsic SRO, which leads to a decrease in the conductivity of SRO thin films.

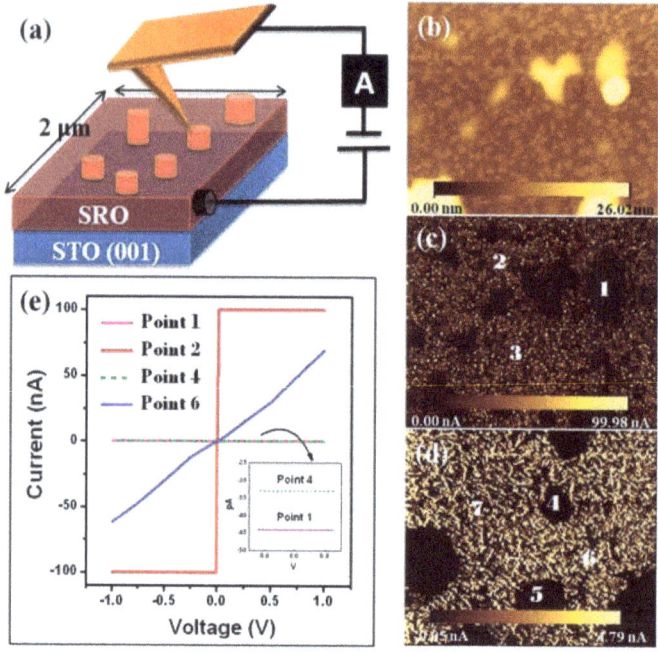

Figure 5. (a) An experimental set-up of a conductive AFM with a scan scale of 2 μm × 2 μm. (b) The surface roughness of Sample C reveals some muffin-like domains. (c) and (d) show the current mapping images for Samples C and D, respectively. (e) The local I-V curves of Points 1, 2, 4, and 6, where the inset figure at the lower and right-hand side is the magnification of the currents of Points 1 and 4, which are on the order of 10–12 Ampere.

4. Conclusions

The growth ambience effects on the structure and conductivity of SRO functional oxide films have been investigated in detail by using a series of deposition oxygen pressures. At a pressure of 10^{-1} Torr, the epitaxial SRO thin film reveals a low resistivity of about 200 μΩ-cm and an excellent crystallization with a sharp FWHM of 0.071° and 0.047° for the x-ray rocking curve and the φ-scan, respectively. The lattice parameter increases gradually below 10^{-4} Torr, and resistivity rises significantly above 10^{-4} Torr as well. The causes are ascribed to the loss in Ru elements and a phase transformation from SRO to $Sr_4Ru_2O_9$. The corresponding microstructures and their conductive properties have also been investigated by utilizing electron microscopy and conductive force microscopy.

Supplementary Materials: The following are available online at http://www.mdpi.com/2079-6412/9/9/589/s1, Figure S1: The growth ambience dependent XRD patterns around surface normal of SRO thin films on STO (001) deposited with the oxygen pressure of 10^{-1}, 10^{-4}, 10^{-6}, and 10^{-7} Torr, respectively, Figure S2. (a) and (b) are X-ray photoemission spectra of Ru $3p$ core-level and O $1s$ for SRO thin film samples, respectively, Figure S3. (a) and

(b) The surface topography and the current mapping images for samples A, respectively. (c), (d), (e) and (f) The corresponding local I-V curves of region A, B, C, and D.

Funding: This research received no external funding.

Acknowledgments: H.-M.C. would like to thank C.Y. Tsai., C. Kao., and S. Yang for operating the PLD system. The input of W.S. Hsu, who provided C-AFM support, is also appreciated.

Conflicts of Interest: The authors declare no conflict of interest.

References

1. Lichtenberg, F. The story of Sr_2RuO_4. *Prog. Solid State Chem.* **2002**, *30*, 103. [CrossRef]
2. Tian, W.; Haeni, J.H.; Schlom, D.G.; Hutchinson, E.; Sheu, B.L.; Rosario, M.M.; Schiffer, P.; Liu, Y.; Zurbuchen, M.A.; Pan, X.Q. Epitaxial growth and magnetic properties of the first five members of the layered $Sr_{n+1}Ru_nO_{3n+1}$ oxide series. *Appl. Phys. Lett.* **2007**, *90*, 022507. [CrossRef]
3. Cao, G.; McCall, S.; Crow, J.E. Observation of itinerant ferromagnetism in layered $Sr_3Ru_2O_7$ single crystals. *Phys. Rev. B* **1997**, *55*, R672–R675. [CrossRef]
4. Klein, L.; Kats, Y.; Marshall, A.F.; Reiner, J.W.; Geballe, T.H.; Beasley, M.R.; Kapitulnik, A. Domain wall resistivity in $SrRuO_3$. *Phys. Rev. Lett.* **2000**, *84*, 6090. [CrossRef] [PubMed]
5. Jung, C.U.; Yamada, H.; Kawasaki, M.; Tokura, Y. Magnetic anisotropy control of $SrRuO_3$ films by tunable epitaxial strain. *Appl. Phys. Lett.* **2004**, *84*, 2590–2592. [CrossRef]
6. Dabrowski, B.; Chmaissem, O.; Klamut, P.W.; Kolesnik, S.; Maxwell, M.; Mais, J.; Ito, Y.; Armstrong, B.D.; Jorgensen, J.D.; Short, S. Reduced ferromagnetic transition temperatures in $SrRu_{1-v}O_3$ perovskites from Ru-site vacancies. *Phys. Rev. B* **2004**, *70*, 14423. [CrossRef]
7. Yoo, Y.Z.; Chmaissem, O.; Kolesnik, S.; Dabrowski, B.; Maxwell, M.; Kimball, C.W.; McAnelly, L.; Haji-Sheikh, M.; Genis, A.P.J. Contribution of oxygen partial pressures investigated over a wide range to $SrRuO_3$ thin-film properties in laser deposition processing. *Appl. Phys.* **2005**, *97*, 103525. [CrossRef]
8. Chang, Y.J.; Kim, C.H.; Phark, S.-H.; Kim, Y.S.; Yu, J.; Noh, T.W. Fundamental thickness limit of itinerant ferromagnetic $SrRuO_3$ thin films. *Phys. Rev. Lett.* **2009**, *103*, 057201. [CrossRef]
9. Xia, J.; Siemons, W.; Koster, G.; Beasley, M.R.; Kapitulni, A. Critical thickness for itinerant ferromagnetism in ultrathin films of $SrRuO_3$. *Phys. Rev. B* **2009**, *79*, 140407. [CrossRef]
10. Grutter, A.; Wong, F.; Arenholz, E.; Liberati, M.; Vailionis, A.; Suzuki, Y. Enhanced magnetism in epitaxial $SrRuO_3$ thin films. *Appl. Phys. Lett.* **2010**, *96*, 082509. [CrossRef]
11. Panchal, G.; Rawat, R.; Bagri, A.; Mandal, A.K.; Choudharya, R.J.; Phase, D.M. Effect of oxygen partial pressure on the electronic and magnetic properties of epitaxial $SrRuO_3$ thin films. *Phys. B Phys. Condens. Matter* **2019**, *572*, 190–194. [CrossRef]
12. Sun, Y.; Zhong, N.; Zhang, Y.-Y.; Qi, R.-J.; Huang, R.; Tang, X.-D.; Yang, P.-X.; Xiang, P.-H.; Duan, C.-G. Structure and electrical properties of epitaxial $SrRuO_3$ thin films controlled by oxygen partial pressure. *J. Appl. Phys.* **2016**, *120*, 235108. [CrossRef]
13. Hiratani, M.; Okazaki, C.; Imagawa, K.; Takagi, K. $SrRuO_3$ thin films grown under reduced oxygen pressure. *Jpn. J. Appl. Phys.* **1996**, *35*, 6212. [CrossRef]
14. Siemons, W.; Koster, G.; Vailionis, A.; Yamamoto, H.; Blank, D.H.A.; Beasley, M.R. Dependence of the electronic structure of $SrRuO_3$ and its degree of correlation on cation off-stoichiometry. *Appl. Phys. Lett.* **2007**, *76*, 075126.
15. Vailionis, A.; Siemons, W.; Koster, G. Room temperature epitaxial stabilization of a tetragonal phase in $ARuO_3$ (A=Ca and Sr) thin films. *Appl. Phys. Lett.* **2008**, *93*, 051909. [CrossRef]
16. Sánchez, A.M.; Äkäslompolo, L.; Qin, Q.H.; Dijken, S.V. Toward all-oxide magnetic tunnel junctions: Epitaxial growth of $SrRuO_3/CoFe_2O_4/La_{2/3}Sr_{1/3}MnO_3$ trilayers. *Cryst. Growth Des.* **2012**, *12*, 954–959. [CrossRef]
17. Velev, J.P.; Duan, C.G.; Burton, D.J.; Smogunov, A.; Niranjan, M.K.; Tosatti, E.; Jaswal, S.S.; Tsymbal, E.Y. Magnetic tunnel junctions with ferroelectric barriers: Prediction of four resistance states from first principles. *Nano Lett.* **2009**, *9*, 427–432. [CrossRef]
18. Herranz, G.; Sánchez, F.; Fontcuberta, J. Domain structure of epitaxial $SrRuO_3$ thin films. *Phys. Rev. B* **2005**, *71*, 174411. [CrossRef]
19. Choi, K.J.; Baek, S.H.; Jang, H.W.; Belenky, J.L.; Lyubchenko, M.; Eom, C.B. Phase-transition temperatures of strained single-crystal $SrRuO_3$ thin films. *Adv. Mater.* **2010**, *22*, 759–762. [CrossRef]

20. Kan, D.; Shimakawa, Y. Strain effect on structural transition in SrRuO$_3$ epitaxial thin films. *Cryst. Growth Des.* **2011**, *11*, 5483–5487. [CrossRef]
21. Singh, D.J.J. Electronic and magnetic properties of the 4d itinerant ferromagnet SrRuO$_3$. *Appl. Phys.* **1996**, *79*, 4818. [CrossRef]
22. Zayak, A.T.; Huang, X.J.; Neaton, B.; Rabe, K.M. Structural, electronic, and magnetic properties of SrRuO$_3$ under epitaxial strain. *Phys. Rev. B* **2006**, *74*, 094104. [CrossRef]
23. Rijnders, G.; Blank, D.H.; Choi, J.; Eom, C.-B. Enhanced surface diffusion through termination conversion during epitaxial SrRuO$_3$ growth. *Appl. Phys. Lett.* **2004**, *84*, 505–507. [CrossRef]
24. Hong, W.; Lee, H.N.; Yoon, M.; Christen, H.M.; Lowndes, D.H.; Suo, Z.; Zhang, Z. Persistent step-flow growth of strained films on vicinal substrates. *Phys. Rev. Lett.* **2005**, *95*, 095501. [CrossRef] [PubMed]
25. Bachelet, R.; Sánchez, F.; Santiso, J.; Fontcuberta, J. Reversible growth-mode transition in SrRuO$_3$ epitaxy. *Appl. Phys. Lett.* **2008**, *93*, 151916. [CrossRef]
26. Estève, D.; Maroutian, T.; Pillard, V.; Lecoeur, P. Step velocity tuning of SrRuO$_3$ step flow growth on SrTiO$_3$. *Phys. Rev. B* **2011**, *83*, 193401. [CrossRef]
27. Yuan, G.L.; Uedono, A. Behavior of oxygen vacancies in BiFeO$_3$/SrRuO$_3$/SrTiO$_3$(100) and DyScO$_3$(100) heterostructures. *Appl. Phys. Lett.* **2009**, *94*, 132905. [CrossRef]
28. Shin, J.; Kalinin, S.V.; Lee, H.N.; Christen, H.M.; Moore, R.G.; Plummer, E.W.; Baddorf, A.P. Surface stability of epitaxial SrRuO$_3$ films. *Surf. Sci.* **2005**, *581*, 118–132. [CrossRef]
29. Lee, H.N.; Christen, H.M.; Chisholm, M.F.; Rouleau, C.M.; Lowndes, D.H. Thermal stability of epitaxial SrRuO$_3$ films as a function of oxygen pressure. *Appl. Phys. Lett.* **2004**, *84*, 4107. [CrossRef]
30. Lee, S.A.; Oh, S.; Hwang, J.-Y.; Choi, M.; Youn, C.; Kim, J.W.; Chang, S.H.; Woo, S.; Bae, J.-S.; Park, S.; et al. Enhanced electrocatalytic activity via phase transitions in strongly correlated SrRuO$_3$ thin films. *Energy Environ. Sci.* **2017**, *10*, 924–930. [CrossRef]
31. Keeble, A.J.; Wicklein, S.; Dittmann, R.; Ravelli, L.; Mackie, R.A.; Egger, W. Identification of A- and B-site cation vacancy defects in perovskite oxide thin films. *Phys. Rev. Lett.* **2010**, *105*, 226102. [CrossRef] [PubMed]
32. Chmielowski, R.; Madigou, V.; Ferrandis, P.; Zalecki, R.; Blicharski, M.; Leroux, C. Ferroelectric Bi$_{3.25}$La$_{0.75}$Ti$_3$O$_{12}$ thin films on a conductive Sr$_4$Ru$_2$O$_9$ electrode obtained by pulsed laser deposition. *Thin Solid Films* **2007**, *515*, 6314–6318. [CrossRef]

 © 2019 by the author. Licensee MDPI, Basel, Switzerland. This article is an open access article distributed under the terms and conditions of the Creative Commons Attribution (CC BY) license (http://creativecommons.org/licenses/by/4.0/).

Review

High Pressure X-ray Diffraction as a Tool for Designing Doped Ceria Thin Films Electrolytes

Sara Massardo [1], Alessandro Cingolani [1] and Cristina Artini [1,2,*]

1. Department of Chemistry and Industrial Chemistry, University of Genova, Via Dodecaneso 31, 16146 Genova, Italy; sara.massardo@edu.unige.it (S.M.); alessandro.cingolani.ac@gmail.com (A.C.)
2. Institute of Condensed Matter Chemistry and Technologies for Energy, National Research Council, CNR-ICMATE, Via De Marini 6, 16149 Genova, Italy
* Correspondence: artini@chimica.unige.it; Tel.: +39-010-353-56082

Abstract: Rare earth-doped ceria thin films are currently thoroughly studied to be used in miniaturized solid oxide cells, memristive devices and gas sensors. The employment in such different application fields derives from the most remarkable property of this material, namely ionic conductivity, occurring through the mobility of oxygen ions above a certain threshold temperature. This feature is in turn limited by the association of defects, which hinders the movement of ions through the lattice. In addition to these issues, ionic conductivity in thin films is dominated by the presence of the film/substrate interface, where a strain can arise as a consequence of lattice mismatch. A tensile strain, in particular, when not released through the occurrence of dislocations, enhances ionic conduction through the reduction of activation energy. Within this complex framework, high pressure X-ray diffraction investigations performed on the bulk material are of great help in estimating the bulk modulus of the material, and hence its compressibility, namely its tolerance toward the application of a compressive/tensile stress. In this review, an overview is given about the correlation between structure and transport properties in rare earth-doped ceria films, and the role of high pressure X-ray diffraction studies in the selection of the most proper compositions for the design of thin films.

Keywords: ionic conductivity; defects association; high pressure X-ray diffraction; doped ceria; thin films; microdevices; solid oxide cells

1. Introduction

Ceria doped by trivalent rare earth (RE) ions forms a wide family of mixed oxides of major relevance for their remarkable properties of ionic conductivity [1]. The partial substitution of Ce^{4+} by RE^{3+} induces in fact the formation of oxygen vacancies which at a sufficiently high temperature are able to migrate through the lattice by hopping from one to another oxygen site [2]. In the presence of an oxygen gradient, such as the one existing between the air and the fuel electrode in a solid oxide cell, the described movement gives rise to an organized and net flow of oxygen ions, thus making RE-doped ceria an ideal class of electrolytes for solid oxide cells. Nevertheless, this is not the only employment of the studied material, being it also used as an ionic conductivity enhancer in mixed ionic electronic conductors (MIECs), namely the perovskite-based materials used for both air and fuel electrodes in solid oxide cells [3]. Moreover, RE-doped ceria is also the active material in many gas sensors [4,5] and memristive systems [6].

The current and increasing need for portable devices makes miniaturization a forefront topic in the materials science research; in this respect, the deposition of thin films having the same or even better properties than the bulk material is one of the main goals connected with the realization of small-scale devices. The issue related to the thin films vs. bulk properties is a crucial point, since the 2D dimensionality, as well as the presence of the film/substrate interface, make the former constitutionally different from the latter, and the properties of thin films unpredictable on the sole basis of the bulk properties.

Citation: Massardo, S.; Cingolani, A.; Artini, C. High Pressure X-ray Diffraction as a Tool for Designing Doped Ceria Thin Films Electrolytes. *Coatings* **2021**, *11*, 724. https://doi.org/10.3390/coatings11060724

Academic Editor: Alberto Palmero

Received: 6 May 2021
Accepted: 12 June 2021
Published: 16 June 2021

Publisher's Note: MDPI stays neutral with regard to jurisdictional claims in published maps and institutional affiliations.

Copyright: © 2021 by the authors. Licensee MDPI, Basel, Switzerland. This article is an open access article distributed under the terms and conditions of the Creative Commons Attribution (CC BY) license (https://creativecommons.org/licenses/by/4.0/).

Ionic conductivity in doped ceria films is well known to be affected by crystallographic parameters, such as the oxide/substrate lattice mismatch, which determine the interfacial strain [7,8], and by the amount and the nature of defects [9]. An effective approach aimed at avoiding the experimental difficulties associated to the study of interfaces consists in the investigation of the high pressure behaviour of bulk samples to simulate the interfacial strain [10] and deduce the material compressibility. Following this idea, recently, several high pressure X-ray diffraction studies were performed by this [11–13] and other research groups [10] with the aim to determine the bulk modulus of several samples belonging to different RE-doped ceria systems, and finally to gain useful hints for the selection of the most proper compositions in the design of thin films.

The review is divided into three main parts. In Section 2, an overview is given about the application areas of RE-doped ceria thin films. Starting from Section 3, the focus moves to the employment of doped ceria in solid oxide cells; this section is devoted to the correlations between structural and transport properties both in bulk and thin films. Finally, in Section 4, a common set-up of high pressure synchrotron X-ray diffraction experiments is described; then, a review of the results obtained from the application of this technique to different bulk doped ceria systems in terms of zero pressure bulk modulus is addressed, and the conclusions to be drawn from this outcome are discussed.

2. Rare Earth-Doped Ceria Thin Films: A Versatile Material to Be Used in Many Fields

As aforementioned, ionic conductivity is the most relevant physical property of RE-doped ceria systems, and the most exploitable one in terms of applications: its maximization, which depends on several different issues, such as composition, RE chemical nature, microstructure, and also extrinsic factors, like the synthesis conditions, plays a central role in driving the research on this material [14]. The occurrence of significant values of ionic conduction, generally ranging between 0.01 and 0.1 S cm^{-1} at 873 K [15], makes doped ceria more desirable in solid oxide cells than yttria-stabilized zirconia (YSZ), which effectively works only at higher temperatures. The employment of doped ceria allows in fact the reduction of the operating temperature of the cell, thus favoring a lengthening of the device lifetime.

The first and foremost employment of RE-doped ceria (in particular for RE ≡ Gd and Sm) is thus in solid oxide cells, either fuel (SOFCs) or electrolysis (SOECs) cells, where it is used mainly as a solid electrolyte [16,17]. To this purpose, it is generally fabricated as a highly dense and thin pellet, in order to favor the movement of oxygen ions through the material. In addition, it is also often mixed, either by co-electrospinning [18] or by infiltration [3,19], to the electrode material with the aim to improve the electrochemical activity of the latter by enhancing its ionic conductivity. In SOFCs, for instance, the presence of Gd- or Sm-doped ceria within the electrode material allows the oxygen reduction reaction (ORR) to take place not only at the triple phase electrolyte/oxygen/air electrode boundary (TPB), but also in the electrode interior [20]. While, in fact, the electrode material is characterized by a high electric conductivity, it has mostly a negligible ionic conductivity, such as in the case of lanthanum strontium manganite (LSM), which is one of the most studied and used cathode materials in SOFCs [21]. Even in the case of lanthanum, strontium, cobalt ferrite (LSCF), presenting a higher ionic conductivity, the addition of some doped ceria significantly improves the material performance [20]. Again, in SOFCs it is also sometimes added to the perovskite fuel electrode material, with the aim to enhance charge transfer under reducing conditions, and in general to improve the electrode morphological stability [3,22]. In Figure 1, taken from [18], the arrangement of a SOFC cathode is shown: the Gd-doped ceria (GDC) electrolyte is in contact with the GDC/LSCF co-electrospun cathode. The EAT (electrochemical active thickness) is the thin layer of electrode where the ORR takes place. The so-formed electrons move through the GDC fibers toward the GDC electrolyte. The gold mesh acts as a current collector, and Pt ensures the contact between gold and the cell.

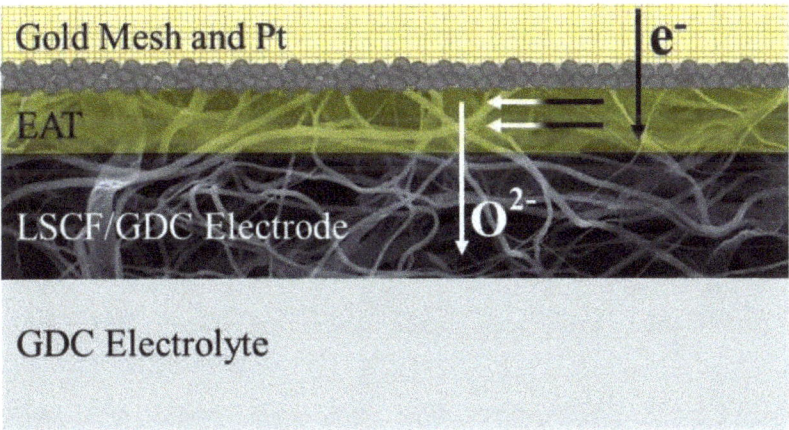

Figure 1. Scheme of a GDC/LSCF co-electrospun electrode. The EAT is the electrochemical active thickness, namely the layer of the electrode where the charge transfer reaction occurs. Reprinted with permission from ref [18]. Copyright 2021 Elsevier License Terms And Conditions.

For sake of completeness, it cannot be overlooked that ceria is also widely studied for its gas sensing properties, thanks to its ability to release or incorporate oxygen as a result of changes in the ambient oxygen concentration [4,5], and to respond to the presence of volatile organic compounds (VOCs) [23]. The partial substitution of Ce^{4+} by a trivalent ion, such as Sm^{3+}, was proven to induce an augmented sensitivity to oxygen of the resulting sensor [24]; the deposition of a Pr-doped ceria film on the yttria-stabilized zirconia-based gas sensor, for instance, is reported to be a valid enhancer of the toluene-sensing properties of the device [25].

A further and more recent application of RE-doped ceria applies to non-volatile memories employing ions in place of electrons as carriers, namely memristive systems [6]. These devices are resistive switches consisting in a switching material interposed between two electrodes; within the cited material a redox reaction takes place, thus giving rise to a resistivity variation. The application of a voltage promotes the flow of O^{2-} ions through the device. Among switching materials, CeO_{2-x} [26] and $Ce_{1-x}Gd_xO_{2-x/2}$ [27,28] are widely studied.

3. Ionic Conductivity in Doped Ceria Bulk and Thin Films: Correlations between Structure, Microstructure and Transport Properties

3.1. Ionic Conductivity in Bulk Doped Ceria

Ionic conductivity in ceria doped by a trivalent RE ion occurs through hopping of oxygen ions toward oxygen unoccupied crystallographic sites; or, in other words, through the movement of oxygen vacancies from one to another oxygen site. The occurrence of empty oxygen positions takes place to ensure electroneutrality as a consequence of the partial replacement of Ce^{4+} by RE^{3+}. The described phenomenon becomes significant above a certain temperature: Sm- and Gd-doped ceria, namely two of the most effective systems in terms of ionic conduction, are commonly used in solid oxide cells working in the intermediate temperature range, i.e., between 673 and 973 K (IT-SOC). The dependence of ionic conduction (σ) on temperature (T) is ruled by the Arrhenius Equation:

$$\sigma T = A exp\left(-\frac{E_a}{RT}\right) \tag{1}$$

where A is the pre-exponential factor, E_a is the activation energy to ionic conductivity, and R the gas constant. E_a can be thought as the sum of association energy (E_{ass}) and migration energy (E_m), being the former the energy needed to separate an oxygen vacancy from a defect aggregate, and the latter the energy needed by vacancies to move through the lattice. Vacancies are in fact free to move only until RE^{3+} ions are incorporated into the CeO_2 crystal structure as isolated RE'_{Ce} defects and oxygen vacancies randomly distributed within the matrix, both acting as guests of a CeO_2-based solid solution [14]. Above a certain RE amount vacancies are partly blocked within defect aggregates.

The reasons behind this behavior can be found in the structural arrangement of both the CeO_2-based solid solution and defect aggregates. CeO_2 crystallizes in a fluorite-like cubic structure (called F) belonging to the $Fm\overline{3}m$ space group [29] containing four formula units per cell; Ce and O are hosted at the two atomic positions $4a$ (0,0,0) and $8c$ (1/4, 1/4, 1/4), respectively, and Ce is eight-coordinated to O. As aforementioned, up to a certain RE^{3+} amount, RE'_{Ce} defects and oxygen vacancies occupy the $4a$ and $8c$ sites respectively, with vacancies being independent of their ionic counterpart and free to move at sufficiently high temperature. With increasing the RE^{3+} content, doping ions and vacancies are stabilized by assuming the cubic atomic arrangement (called C) of the RE_2O_3 sesquioxides of the heaviest rare earths (RE ≡ Gd-Lu) (space group: $Ia\overline{3}$), where RE is six-coordinated to O [30]; in this configuration, vacancies cannot flow through the lattice, unless E_{ass} is provided to the system in order to dislodge them from defect clusters. In Figure 2 the atomic arrangements of the F and the C phase are represented.

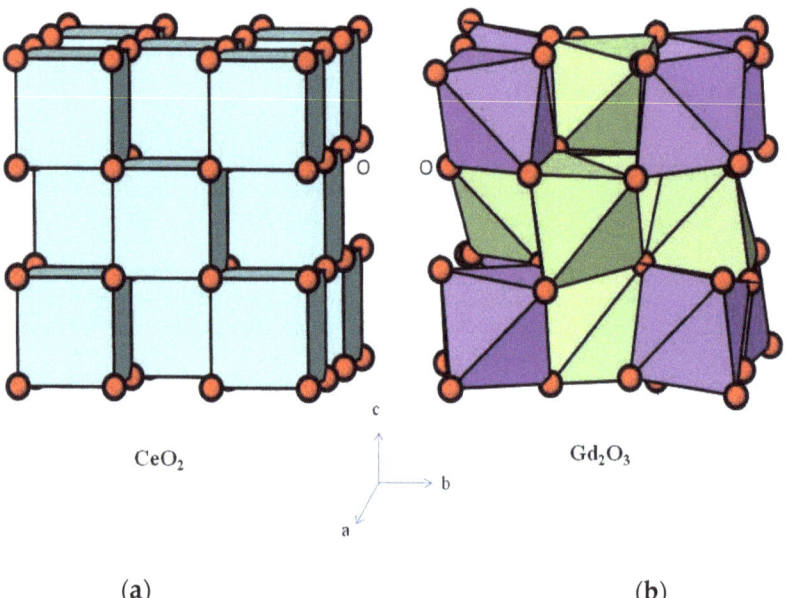

(a)   (b)

Figure 2. Crystal structures of (**a**) CeO_2 (F phase) and (**b**) Gd_2O_3 (C phase). To make the comparison easier, only one eighth of the Gd_2O_3 unit cell is drawn. The lanthanide element is located at the center of each polyhedron. Reprinted with permission from ref [31]. Copyright 2021 Elsevier License Terms And Conditions.

From this perspective, it is possible to understand why in many doped ceria systems two different values of activation energy can be recognized from the Arrhenius plot, with the lower one at higher temperature [32–35]. This is quite a common feature in doped ceria systems, and it is ascribable to the occurrence of at least two different defect aggregates, namely $1V_O^{\bullet\bullet}RE'_{Ce}$ positively charged dimers and $1V_O^{\bullet\bullet}2RE'_{Ce}$ neutral trimers [36]. In general, a higher binding energy is calculated for dimers rather than for trimers [37]: as a consequence, the latter, due to their lower configurational entropy, become progressively less stable with increasing temperature, and their dissociation at a sufficiently high temperature induces the release of oxygen vacancies which become able to migrate, thus reducing the activation energy value. The crossover temperature results to be roughly the same for all the systems, and it was observed around 750 K [34]; in Figure 3, reproduced from [34], the trend of the activation energy below and above 750 K, as well as the pre-exponential factor A appearing in Equation (1), are reported as a function of the ionic size of the doping ion.

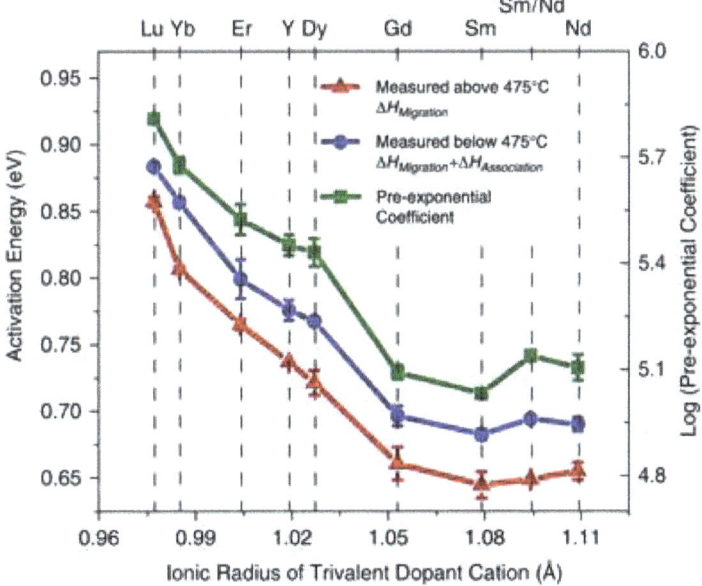

Figure 3. Behaviour of the activation energy E_a below and above 750 K, and of the pre-exponential factor A as a function of the ionic size of the doping ion. Reprinted with permission from ref [34]. Copyright 2021 John Wiley and Sons License Terms and Conditions.

In spite of the large differences in ionic conductivity existing between systems doped with different RE^{3+} [34], the maximum in σ (σ_{max}) occurs within a narrow compositional interval, namely at x (in $Ce_{1-x}RE_xO_{2-x/2}$) ranging between 0.10 and 0.25 for the majority of the studied systems, such as for $RE \equiv Gd$ [2], Sm [38], Lu [39], as well as (Nd,Tm) [33]. At higher x, ionic conductivity significantly drops, and its absolute value becomes no more significant for any application; nevertheless, even heavily RE-doped systems are of great interest in fundamental research, since they allow to correlate subtle structural details to transport properties [31,32,40–42]. The substantial coincidence of the position of σ_{max} in different systems suggests that the appearance in diffraction patterns of peaks attributable to defect clusters at largely different x values for different RE ions, does not mean that the F phase has a different extent according to the RE identity; on the contrary, C defect aggregates are present and stable even at very low RE concentration, but due to their different spatial correlation, they can be identified by X-ray diffraction at different x values [33].

In addition to the numerous studies performed on singly-doped ceria systems, also a large number of doubly- and multiply-doped systems has been investigated, such as Gd/Y- [43], Gd/Sm- [44], La/Sm- [35], Sm/Nd- [45], Nd/Gd- [46], La/Dy- [47], Gd/Sm/La [48]. A general lowering of activation energy [33] and an improvement of the ionic conductivity properties [43,49,50] with respect to singly-doped ceria is often observed.

3.2. Ionic Conductivity in Doped Ceria Thin Films

When approaching thin films, the first and foremost character to be taken into account is their 2D nature, which strongly affects all their properties. Going into detail of the particular case of doped ceria, ionic conductivity results to be specifically driven, in addition to the factors mentioned in the previous paragraph, also valid for the material in bulk form, by the strain arising at the film/substrate interface, and by the dislocations content.

The unavoidable lattice mismatch occurring at the film/substrate interface is responsible for the former item, namely for the occurrence of strain. Tensile strain, in particular, is believed to operate a decrease in the activation energy to ionic conduction, and therefore an improvement of this property, while the opposite occurs when compressive strain arises [51,52]. This happens because a migrating oxygen ion, while hopping from one to another vacancy site, has to move through a narrow aperture, namely the space between two cations, which is generally much smaller than the oxygen ion radius. A tensile strain contributes to widen this aperture, thus favoring the movement of oxygen ions. As the activation energy to ionic conduction is directly correlated to the lattice perturbation needed to let the O^{2-} ion through, a tensile strain helps to reduce this quantity. This evidence is further corroborated by the increase in the activation energy observed in buckled free-standing doped ceria membranes, which are subjected to a compressive stress due to the buckling itself [53]. For instance, an increase of 0.13 eV was observed in the activation energy of a $Ce_{0.8}Gd_{0.2}O_{1.9}$ free-standing membrane as a response to a 1.78% compressive strain with respect to the flat substrate-supported thin film with the same composition [54]. Similarly, a multilayer oxide-based device consisting in alternating layers of $Ce_{0.9}Gd_{0.1}O_{1.95}$ and Er_2O_3, where the number of individual layers was changed from 1 to 60 in order to vary the strain-activated volume, presents an increase in activation energy by 0.31 eV with enhancing the compressive strain by 1.16% [54].

The effect of strain on ionic conductivity is comparable to that of introducing doping ions into the ceria lattice: inducing strain by using a doping ion with a significantly different size than Ce^{4+} causes in fact a change in ionic conduction, due to the variation of internal pressure caused by the foreign ion [34]. Computational simulations performed on $Ce_{1-x}Y_xO_{2-x/2}$, for example, suggest that the application of isotropic strain to the bulk material induces an increase in ionic conductivity by three times at 1273 K for $x = 0.18$ [8]. It can be thus concluded that both doping and the existence of a lattice mismatch have to be carefully considered when designing doped ceria thin films. Therefore, the desired transport properties can be engineered as a combined result of internal and external strain.

In light of the previous considerations, the aforementioned film/substrate lattice mismatch comes as an ideal item, acting as a strain source; coherently grown films, in particular, are especially affected. Nonetheless, provided that the lattice mismatch, if exerting a tensile strain, plays a positive role in improving ionic conductivity, it should never overcome a certain threshold, which primarily depends on the elastic properties of the material; otherwise, the excess strain is released through the formation of dislocations. The segregation of both trivalent cations and oxygen vacancies in the vicinity of the dislocation satisfies the natural tendency to reduction of elastic energy around these defects, and it necessarily reduces ionic conductivity [55]. For this reason, a direct growth of doped ceria on a MgO substrate is not recommended: the 8.62% lattice misfit ε ($\varepsilon = \frac{a_{ceria} - a_{substrate}}{a_{ceria}} \cdot 100$, with a the lattice parameter) is too large to ensure a proper substrate/film match [56]. If, on the contrary, a $SrTiO_3$ buffer layer is interposed between MgO and the film, in the particular case of Sm-doped ceria ε is reduced to 1.62% if the cubic axis of the growing film is rotated by 45° with respect to the one of the substrate, thus allowing to take advantage of a significant ionic conductivity increase. Moreover, the lattice mismatch lies at the root of the often observed columnar growth of the film, which takes place as an attempt to accommodate interfacial strain [57].

That being said, the literature provides a large amount of even not entirely agreeing data; nevertheless, even if the colossal ionic conductivity increase by a factor 10^8 observed in YSZ grown on $SrTiO_3$ and ascribed to the very large tensile strain [58] does not take place in doped ceria, most studies point at an increase in ionic conductivity of strained films with respect to their bulk unstrained counterpart. This evidence is generally deemed as a direct consequence of the increase of mobile ions at the strained interface [59]. $Ce_{0.8}Gd_{0.2}O_{1.9}$ thin films grown on a MgO crystal, for instance, are reported to present very similar values of ionic conductivity and activation energies as the corresponding bulk material, perhaps due to the very large lattice mismatch [60], while $Ce_{0.8}Sm_{0.2}O_{1.9}$ films grown on $SrTiO_3$-buffered MgO present at 973 K a σ value of 0.07 S·cm^{-1}, to be compared to σ = 0.02 S·cm^{-1} deriving from dense polycrystalline pellets [57] an even larger ionic conductivity enhancement takes place in Sm-doped ceria/YSZ films deposited on the same substrate [60]. A clear increase in ionic conductivity with increasing strain was observed at each temperature considered in the range 723–1123 K in epitaxial $Ce_{0.9}Gd_{0.1}O_{0.95}$ thin films grown on $SrTiO_3$-buffered MgO [7].

It is the interplay between two main parameters which rules the dependence of ionic conductivity on strain: film thickness and grain size. The former can be understood in terms of volume fraction of strained material: the thinner the film, the larger the strained volume fraction. A strong increase in ionic conductivity for instance was observed in epitaxial grown Sm-doped ceria/YSZ films with reducing film thickness [61]; moreover, an ideal film thickness of ~80 nm was suggested to achieve the maximum ionic conductivity in $Ce_{0.8}Gd_{0.2}O_{0.9}$ films deposited by pulsed laser deposition on single crystal (0001) Al_2O_3 substrates, as the best compromise between the strain induced by lattice misfit and the grain boundary conduction associated to grain size [62]. Nevertheless, even the thickness of the buffer layer plays a role: a controlled reduction of the $SrTiO_3$ layer implies a weaker compensation of the ceria/MgO lattice mismatch, thus leading to an enhancement of ionic conductivity, as experimentally observed in epitaxial $Ce_{0.9}Gd_{0.1}O_{0.95}$ thin films [63]. Finally, it is noteworthy that an increase in ionic conductivity is also observed in doubly-layered Sm-doped ceria/YSZ films with increasing the amount of bilayers and keeping constant the overall film thickness [62].

As aforementioned, even the grain size is a relevant item when discussing the ionic conductivity of doped ceria films. Generally speaking, the reduction of the grain size always implies an increase in the grain boundary volume and, as a consequence, a progressively larger contribution to transport properties of the latter. This normally translates into a conductivity decrease, at least in bulk materials. In ionic conducting thin films, something different happens. As already observed and discussed by Christie and Berkel [63], in such films a grain size decrease induces an increase in the grain boundary conductivity. An increase in σ, for example, was observed in $Ce_{0.8}Gd_{0.2}O_{0.9}$ films deposited by spin-coating on a sapphire substrate with decreasing the grain size, most probably due to the reduction of the activation energy as a consequence of the segregation of impurities within the grain boundary volume [64,65]. This is a general achievement; despite the scatter among absolute conductivity values reported in different papers [66–68], it seems to be correlated to the decrease in activation energy, occurring irrespective of the different deposition techniques. Anyway, since films containing smaller grains display larger strain [66], the described evidence is a further clear indication toward a direct correlation between strain and ionic conductivity.

3.3. Thin Films Deposition Processes

Deposition techniques useful to prepare doped ceria thin films can be divided into two main groups, namely the vacuum- and the precipitation-based ones. To the former belong all techniques making use of a target consisting in a sintered pellet; the latter, on the contrary, rely on the employment of solvents.

To obtain dense thin films of doped ceria, sputtering and pulsed laser deposition (PLD) are widely used, since they allow to obtain thin doped ceria layers, starting from the precursors powders. For instance, Liu et al. [26] used the rf magnetron sputtering method to deposit a ~16 ± 2 nm film on a Pt/Ti/SiO2/Si substrate for the use in memristive devices, working at room temperature, and starting from a CeO_{2-x} ceramic target, under ultra-fine

Ar atmosphere. Recently, Kumar et al. [16] employed pulsed laser deposition to deposit Sm-Gd co-doped ceria films on Si (111), starting from a dense sintered pellet of Gd_2O_3, Sm_2O_3 and CeO_2 in the molar ratio 1:1:8; the deposition process was performed under different oxygen pressure conditions, to determine the effect of pO_2 on the structural and morphological properties of the grown films.

However, the sputtering and PLD methods are generally not so cost-effective, since they require very complex apparatuses, therefore making the scaling-up of the thin films production process very difficult to achieve. Schlupp et al. [69] proposed an efficient and cheap way to obtain Gd-doped ceria films by aereosol assisted chemical vapor deposition, starting from stoichiometrically mixed precursors solutions of cerium (IV) and gadolinium (III) 2,2,6,6-tetramethylheptanedionate, and performing the deposition on Si (100) and sapphire single crystals. Lair et al. [70] carried out a study on the structural and morphological properties of electrodeposited samaria-doped ceria films, to determine their possible use in μ-SOFCs. In the mentioned experiment, thin films were deposited on a stainless steel substrate at room temperature (303 K), from aqueous electrolytic solutions in which a supporting electrolyte ($NaNO_3$ 0.1 M) and stoichiometric amounts of $Ce(NO_3)_3 \cdot 6H_2O$ and $Sm(NO_3)_3 \cdot 6H_2O$ were dissolved.

In this section, only a few possible techniques for the ceria-based thin films preparation are presented, but several others can be found in literature [71]: a more detailed discussion is beyond the aim of the present paper.

4. High Pressure Diffraction Studies on Bulk Doped Ceria: A Bridge between Bulk and Thin Films

4.1. Why High Pressure X-ray Diffraction?

X-ray diffraction is well known to be a powerful experimental technique, which provides a fundamental contribution to the comprehension of structure [41], microstructure [72], occurrence of defects [13], and crystallinity [73]. In the particular case of doped ceria, it is evident that strain contributes in a fundamental way to ionic conduction of thin films. As aforementioned, tensile strain has to be high enough to enable a significant reduction of the activation energy to ionic conduction, but at the same time it should not exceed a certain threshold, in order to avoid the appearance of dislocations. In this respect, thin film compositions able to accommodate a large strain are potential promising candidates ionic conductors. The ability of a material to accommodate strain is expressed by the compressibility of the material (k), namely its ability to modify its volume as a response to a pressure change. Its reciprocal, the isothermal bulk modulus B_0, is defined as:

$$B_0 = -V_0 \left(\frac{\partial P}{\partial V} \right)_{P=0} \quad (2)$$

where V_0 is the cell volume at ambient conditions. If, as it is desirable, k is large, B_0 obviously assumes a small value. The latter condition is also advantageous because it implies a small volume elastic energy, and a substantial suppression of the dislocations generation.

The investigation of strain and strain-related phenomena in thin films is experimentally quite difficult due to the need to exclude the contribution of the substrate from the measurement of any property. Within this framework, in-situ high pressure X-ray diffraction applied to bulk samples of doped ceria comes to aid, allowing to determine for each composition the isothermal bulk modulus B_0 by properly treating the refined cell volumes, and hence k. In this work, the calculation is conveniently limited to the F phase, since this is the atomic arrangement where ionic conduction takes place, but it could be in principle extended even to C. In order to gain an idea of the correspondence between applied pressure and strain, it is interesting to notice that applying a pressure of ~20 GPa to a Gd-doped ceria bulk sample causes a compressive strain of around 5%, while the application of 5 GPa roughly corresponds to a 1% compressive strain [7].

4.2. Fundamentals on the Experimental Method

Diamond anvil cells (DACs) are generally used to apply pressure on a selected sample by pressing it between two perfectly aligned diamond culets that slowly move closer to each other thanks to a gear-driven mechanism: the maximum achievable pressures can vary a lot, mainly depending on the DAC model. The present devices can be used to apply pressure on a given sample either anisotropically or isotropically. In the former case, uniaxial pressure is applied to the sample simply pressing it between the diamonds; in the latter, sample powders are loaded inside the pressure chamber of a specific metallic sample-holder (the so-called gasket), and a pressure transmitting medium (PTM) is required to uniformly distribute pressure within the chamber. Silicon oil, NaCl powders and Ar (in membrane-DACs) are examples of widely used PTM: to ensure the hydrostatic distribution of pressure inside the chamber, an ideal sample:PTM volume ratio of 20:80 is generally needed. In any case, in order to determine the effective pressure exerted on the sample, an internal standard is required: generally, a material with a well known high-pressure structural behaviour (e.g., Cu).

Electromagnetic waves enter the cell passing through the diamonds, and interact with the pressed sample: therefore, the described device is potentially compatible with many experimental techniques, mainly depending on the DAC model and characteristics (e.g., DAC with ultra low fluorescence diamonds for Raman spectroscopy).

As aforementioned, the maximum achievable pressure is strongly influenced by the diamond culet size employed: the lower the culet diameter, the higher the maximum pressure. For instance, a cell with Boehler-Almax designed diamonds with a $\varnothing > 1$ mm culet allows to reach a maximum pressure of 5 GPa, meanwhile in principle, with a $\varnothing = 200$ µm diamond, pressures higher than 100 GPa can be reached. However, also the gasket preparation process must be performed with the utmost care. In fact, the sample chamber consists in a hole drilled throughout a metallic disc that has been previously thinned by being pressed between the diamonds of the cell: an excessive gasket thickness, a non-circular hole or a hole which is not drilled at the centre of the diamond culets impression, or even an overfilled sample chamber can affect the maximum achievable pressure. Moreover, the hydrostatic limit of the selected PTM is also very important: for instance, a 4:1 mixture by volume of methanol and ethanol, which is one of the most commonly used liquid PTM, allows to work in hydrostatic conditions up to 10 GPa, while argon allows to reach higher pressures, since it causes a very small pressure gradient (<1.5%) at 80 GPa [74]. The silicon oil employed in the experiments by the present research group ensured the hydrostatic working conditions up to 10 GPa, therefore covering the whole considered pressure range.

The authors' research group focused on RE-doped ceria systems containing RE $\equiv$ Lu [11], Sm [12] and a mixture of Nd/Tm ions [13]. In particular, the last-mentioned system was synthesized using a Nd/Tm = 0.74/0.26 ratio, in order to reproduce the average ionic size of Sm^{3+} (C.N. 8), being Sm-doped ceria one of the most promising electrolytes for solide oxides cells [75]: $Ce_{0.8}Sm_{0.2}O_{1.90}$, for example, reaches the remarkable σ value of 10^{-2} S cm^{-1} at 773 K [38]. All the samples were prepared by coprecipitation of the corresponding mixed oxalates, and by subsequent thermal treatment at 1173 K for 3 days in order to obtain a high crystallinity degree, as thoroughly described in [41,76]. Going into detail, authors studied $Ce_{1-x}RE_xO_{2-x/2}$ systems with x ranging between 0.1 and 0.4 for RE $\equiv$ Lu [11], 0.2 and 0.6 for RE $\equiv$ Sm [12], and 0.1 and 0.6 RE $\equiv$ (Nd,Tm) [13]. Moreover, regarding the Lu-doped ceria system, even the composition having $x = 0.8$ was studied. Samples are named according to the RE doping ion identity and to its percentage amount with respect to the total lanthanide content. For example, the name Sm20 corresponds to the composition $Ce_{0.8}Sm_{0.2}O_{1.9}$.

High pressure X-ray diffraction data of the aforementioned systems were collected at the Xpress beamline of the Elettra synchrotron facility, located in Basovizza (Trieste, Italy). Pressure was applied to all the doped ceria powders up to ~7 GPa, thus obtaining comparable data for all the considered systems; at least six different pressure values were taken into account for each sample. Powders were loaded in a gear-driven Boehler-Almax plate DAC equipped with diamonds with a 300, 500 or 600 µm culet size (respectively for RE ≡ Nd/Tm, Sm and Lu), and characterized by a large X-ray aperture. For instance, the 300 µm culets DAC is characterized by an angular aperture of 60°, thus allowing to collect the image of the diffraction rings in the angular range from −30° to 30°. The pressure chambers were settled by drilling hole with a diameter of 100, 200 or 150 µm (respectively for RE ≡ Nd/Tm, Sm and Lu) through spark-erosion in several 200-µm-thick rhenium gaskets, which had been previously indented to reduce their thickness below 110 µm (70 for RE ≡ Lu, for which stainless steel gaskets were employed). Silicone oil was used as the pressure transmitting medium, and pressure calibration was performed by two different methods: either by placing Cu foils into the sample chamber (as depicted in Figure 4), and taking into account the displacement of the (111) diffraction peak with increasing pressure, or by using ruby chips, studying the displacement of their fluorescence lines. All the diffraction analyses were performed in hydrostatic conditions.

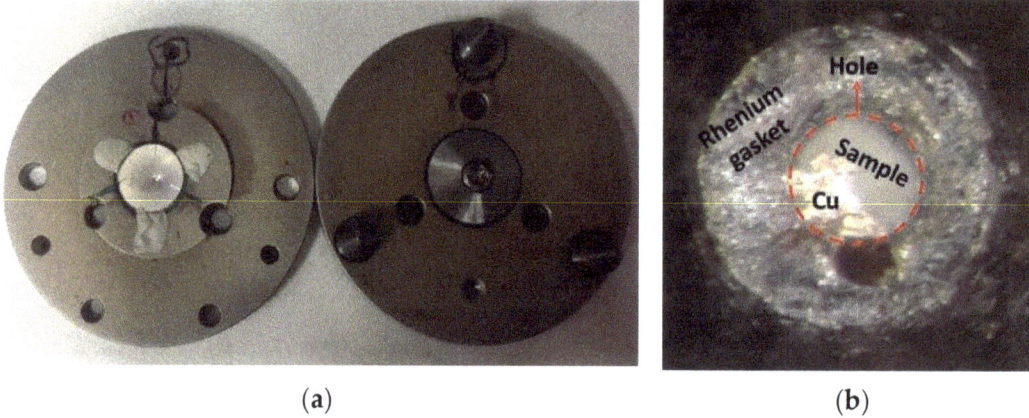

Figure 4. (a) Lower (on the left) and upper (on the right) plate of the 300 µm DAC by Boehler-Almax. The upper diamond is observable, meanwhile the lower one is covered by the gasket, which is fixed onto its holder by using some play dough. (b) The pressure chamber filled with the sample powders. The internal standard (Cu) and the PTM are present.

The Xpress beamline is equipped with a MAR345 image plate detector allowing to collect images of the diffraction rings, which are subsequently converted into the typical 2ϑ-intensity diffraction patterns by the Fit2D software [77]. The 2ϑ range spanned between 5° and 24° for RE ≡ Nd/Tm, between 6° and 36° for RE ≡ Lu, and between 4° and 30° for RE ≡ Sm. The FullProf Suite [78] was used to refine by the Rietveld method [79] the structural models reported in Section 3.1, which describe the obtained diffraction patterns.

4.3. Fundamentals on Data Analysis

Diffractograms were treated through the Rietveld method by refinement of the structural models which better describe experimental data. As elucidated in Section 3.1, RE-doped ceria crystallizes in the F structure up to a certain RE^{3+} doping ion amount; beyond this limit, defect aggregates characterized by the C structure appear. If Ce^{4+} and RE^{3+} are sufficiently close in size, the cited C defect clusters act as guests of the CeO_2-based solid solution. Since C is a superstructure of F, and due to the aforementioned Ce^{4+}/RE^{3+} close size similarity, such as in the case of RE ≡ Gd^{3+} and Sm^{3+} (Ce^{4+}, CN:8, r = 0.97 Å; Sm^{3+}, CN:6, r = 0.958 Å; Gd^{3+}, CN:6, r = 0.938 Å [80]), diffraction patterns of C and F only differ by the superstructure peaks, while all the other ones coincide. The exactly double length of the C cell parameter with respect to the one of F is a further proof of that. As the amount of C defect aggregates gradually grows with increasing the RE^{3+} content, neither the F nor the

C model properly describe the real atomic arrangement; on the contrary, an intermediate or hybrid model (called H), is the best candidate. In H, structural parameters sensitive to the F/C passage, such as the x coordinate of the 24d site, occupied by Ce and RE, and the occupancy factor of the 16c site, populated by O, display a progressive transition from the values typical of F to the ones of C with increasing the RE content. If, on the contrary, Ce^{4+} and RE^{3+} are too different in size, namely starting from RE ≡ Tm and going toward smaller RE elements [81], a two-phase field (F + C) becomes stable beyond the stability region of the CeO_2-based solid solution. In the light of these considerations, among the studied systems, while Sm- and (Nd,Tm)-doped ceria show the hybrid region, in Lu-doped ceria the biphasic (F + C) field appears. Relying on the occurrence of diffraction peaks, in $Ce_{1-x}RE_xO_{2-x/2}$ the F region extends up to x = 0.3, 0.4 and 0.5 for RE ≡ Sm, Lu, and (Nd,Tm), respectively. This means for instance that in sample Lu80 a mixture of the F and C phase is present, with C largely prevailing over F [42]. Structural models were refined accordingly; for sake of clarity, the F, C and H structural models are reported in the Supplementary Materials (Table S1).

In order to elaborate each diffraction pattern, the peak profile was optimized by the pseudo-Voigt function, while the background was refined by linear interpolation of a set of ~70 experimental points. In the last refinement cycle, structural parameters (cell parameters, and for the H phase the Ce/RE1 x coordinate, the O1 x, y and z coordinates, and both the occupancy factor and the x coordinate of O2), atomic displacement parameters, the scale factor, five peak parameters and the background points were allowed to vary. For each sample, high pressure data were refined using as starting parameters the optimized values of atomic positions found at ambient pressure. In compositions containing Cu as internal standard, angular regions where peaks of Cu occur were excluded from refinements. In Figure 5, the Rietveld plot of sample Sm60 at 4.11 GPa is reported as a representative example. The diffraction pattern presents the so called hybrid structure (H), resulting from the intimate intergrowth of the F and the C lattices (see Section 4.3), where C is a superstructure of F. The main peaks detectable in the diffraction pattern are the ones common to F and C; the minor peaks, on the contrary, can be ascribed to the C phase contribution.

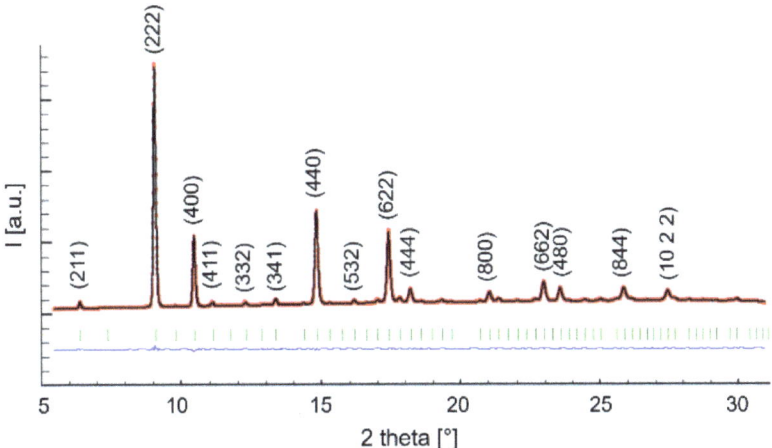

Figure 5. Rietveld refinement plot of sample Sm60_4.11. The red and the black lines are the experimental and the calculated diffraction pattern, respectively; the blue line is the difference curve; vertical bars indicate the calculated positions of Bragg peaks.

Performing high pressure X-ray diffraction analyses on a given system allows to determine the equation of state (EoS) by studying the evolution of the refined unit cell volumes under the effect of an applied pressure. The isothermal EoS of a solid is quite complex, and it is generally satisfactorily described by a refinable model; among all the possibilities, the four most common equations are Murnaghan [82], Birch-Murnaghan [83] natural strain [84], and Rose-Vinet [85]. All of them express the relation between the unit cell volume of the material at ambient conditions (V_0) and pressure (P), in terms of bulk modulus at zero applied pressure and its first pressure derivative (B_0 and B'_0, respectively). The Murnaghan EoS [82], in particular, relies on the assumption that the bulk modulus linearly varies with pressure; according to it, the dependence of pressure on volume can be written as:

$$P(V) = \frac{B_0}{B'_0}\left[\left(\frac{V_0}{V}\right)^{B'_0} - 1\right] \quad (3)$$

A further development of this theory is represented by the work of Birch [83], who concentrated on the application to a cubic cell of strain consisting in a hydrostatic compression plus a homogeneous infinitesimal strain. The treatment gives rise to the following form of $P(V)$:

$$P(V) = \frac{3B_0}{2}\left[\left(\frac{V_0}{V}\right)^{\frac{7}{3}} - \left(\frac{V_0}{V}\right)^{\frac{5}{3}}\right]\left\{1 + \frac{3}{4}(B'_0 - 4)\left[\left(\frac{V_0}{V}\right)^{\frac{2}{3}} - 1\right]\right\} \quad (4)$$

The Natural Strain EoS, which was elaborated by Poirier and Tarantola [84], is a logarithmic equation expected to be valid for a wider pressure range than the aforementioned ones, and it can be expressed as follows:

$$P(V) = B_0 \frac{V_0}{V}\left[\ln\left(\frac{V_0}{V}\right) + \frac{(B'_0 - 2)}{2}\left[\ln\left(\frac{V_0}{V}\right)^2\right]\right] \quad (5)$$

Finally, the Rose-Vinet EoS [85] was originally thought to predict the high temperature response of a material to the application of an external pressure; to accomplish this goal, the knowledge at each different temperature of B_0, B'_0, V_0 and of the coefficient of thermal expansion (CTE) at zero pressure, are needed. Following the proposed approach, it is possible to provide an estimation of the temperature dependence of CTE, B_0 and B'_0. The expression of $P(V)$ follows.

$$P(V) = \frac{3B_0}{X^2}(1-X)exp[\eta_0(1-X)] \quad (6)$$

with

$$X = \left(\frac{V}{V_0}\right)^{1/3} \quad (7)$$

and

$$\eta_0 = \frac{3}{2}[B'_0 - 1] \quad (8)$$

Starting from the refined unit cell volumes of the F phase, the most suitable EoS was determined for each sample by using the EoSFit7_GUI software [86]. In particular, for all the considered systems, different EoS were tested, and in spite of the generally negligible discrepancies among the results deriving from different equations, the best results were always reached using the third order Rose-Vinet equation.

Refinements of the equation of state through the described Rose-Vinet model provided the results reported in Table 1.

Table 1. B_0 and B'_0 values obtained from the refinement of the third order Rose-Vinet equation of state applied to the three studied systems.

Sample	B_0[GPa]	B'_0
Sm20	258(31)	−2(2)
Sm30	243(23)	23(2)
Sm40	211(40)	68(6)
Sm50	156(29)	39(2)
Sm60	169(51)	46(2)
Lu10	172(18)	24(6)
Lu20	203(76)	13(2)
Lu30	191(15)	7(1)
Lu40	170(9)	5(3)
Lu80	73(10)	85(4)
NdTm10	290(32)	34(14)
NdTm20	260(35)	18(3)
NdTm30	194(20)	31(12)
NdTm40	107(11)	121(17)
NdTm50	161(32)	7(9)
NdTm60	186(34)	37(16)

4.4. High Pressure X-ray Diffraction Studies on Sm-, Lu-, and (Nd,Tm)-Doped Ceria: How They Can Drive the Design of Thin Films

The examination of data reported in Table 1 reveals in each system a generally decreasing trend of B_0 of the F phase with increasing the RE content. This finding is in good agreement with the results presented by Rainwater et al. [10] on Sm-doped ceria: notwithstanding the slight different absolute B_0 values, a significant decrease of bulk modulus with increasing the doping ion content occurs even in their data.

The described behavior can be interpreted in the light of the masterly work by Anderson and Nafe [87], who in 1965 studied the correlation between bulk modulus and mean atomic volume (V_{at}) of covalent crystals and oxides. They moved from the observation that $\ln B_0$ shows a linear correlation to $\ln(2V_{at})$, which is obeyed by a huge variety of compounds, and finally accomplished the development of the following empirical law:

$$\ln B_0 = -m \ln(2V_{at}) + constant \qquad (9)$$

Equation (9) is valid both for covalent crystals and oxides, with the only difference that m assumes the value 4/3 in the former compounds, while it ranges between 3 and 4 in the latter. It suggests that oxides are characterized by a much stronger decrease of B_0 with increasing V_{at} with respect to covalent compounds.

In Figure 6, the trend of $\ln B_0$ vs. $\ln(2V_{at})$ is reported for all the studied systems, together with the line representing the expected trend for oxides according to Anderson and Nafe; the latter was built attributing to m the intermediate value 3.5. At first sight it can be observed that qualitatively even doped ceria follows the trend common to oxides. V_{at} is defined as the lattice volume divided by the number of atoms therein contained; it is, therefore, noteworthy that with increasing the content of the trivalent doping ion, the oxygen amount, and hence the total number of atoms per formula unit, proportionally decreases. It results thus clear that within the considered systems, an increase in V_{at} derives from an increase in the RE^{3+} content. A closer observation of the diagram reveals that, even if the trend of doped ceria systems roughly complies with that of other oxides, the decreasing rate is higher for the former, being for instance m ~6 for both Sm- and (Nd,Tm)-doped ceria at low doping content. Moreover, it can also be observed that the deviation from the expected trend becomes progressively larger on going from Sm- to (Nd,Tm)- to Lu-containing systems, i.e., with introducing progressively smaller doping ions. This evidence finds an explanation considering that the binding energy, and thus the stability, of C defect aggregates becomes larger with decreasing the RE^{3+} size [37]; therefore, at a given

x in $Ce_{1-x}RE_xO_{2-x/2}$, for $RE \equiv Lu$ the amount of C defect associates is higher than for $RE \equiv (Nd,Tm)$ and Sm. The case of doubly-doped ceria is particularly intriguing, because out of Nd^{3+} and Tm^{3+}, the latter preferentially and in large amount enters C defect clusters, being Tm^{3+} the smaller of the two ions. This item leads the (Nd,Tm)-doped system to present lower B_0 values than the Sm-containing system, notwithstanding the identical average ionic size of doping ions in the two systems.

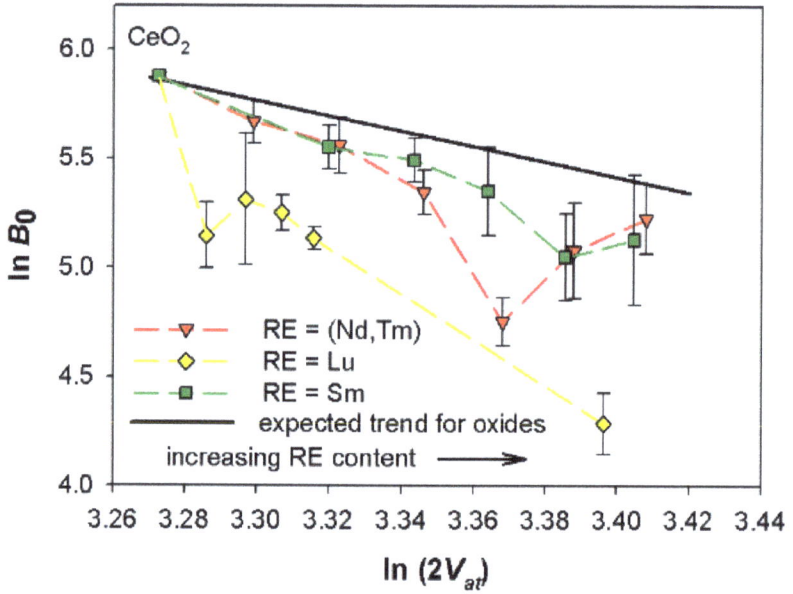

Figure 6. Trend of lnB_0 vs. $ln(2V_{at})$ for all the studied systems. Data of CeO_2 are taken from [88], while the ones of the Sm- and (Nd,Tm)-containing systems derive from [13]; experimental points of Lu-doped ceria are taken from [12]. Cell volumes of samples crystallizing in the F phase are multiplied by 8 in order to make them comparable with those crystallizing in the C structure.

The subtraction of RE ions from the F matrix through the formation of C domains implies that the total number of atoms/unit formula of the F phase is overestimated if the nominal stoichiometry of the overall oxide is taken into consideration. This issue explains the discrepancy between the expected and the experimental trend of $\ln B_0$ vs. $\ln(2V_{at})$, and it is thoroughly discussed in [13], where an estimate of the defect clusters amount is also provided.

The significant difference in B_0 between different systems at a given x, clearly visible in Figure 3, is also revealed in compressibility. Figure 7 depicts the behavior of compressibility k as a function of the doping ion amount x. A general increasing trend can be appreciated with increasing x for each system, but even more interestingly, the presence of smaller doping ions promotes the occurrence of larger compressibility values at each given x.

These findings have some important implications in the framework of the design of thin films. As previously discussed, ionic conductivity in thin films is strictly dependent, in addition to other factors, also on the amount of tensile interfacial strain which can be accommodated without releasing it through the formation of dislocations; in other words, it strongly depends on the material compressibility. Therefore, the results presented in this work suggest that, in terms of compressibility, the employment of (a) small doping ions and (b) high doping contents are to be preferred. These guidelines need of course to be reconciled with other parameters contributing to ionic conduction, such as the effect of the RE chemical identity, and the observed drop beyond a certain x value. Nevertheless, they can be considered as a further reference point to be taken into account when designing doped ceria thin films to be used as electrolytes in solid oxide microdevices.

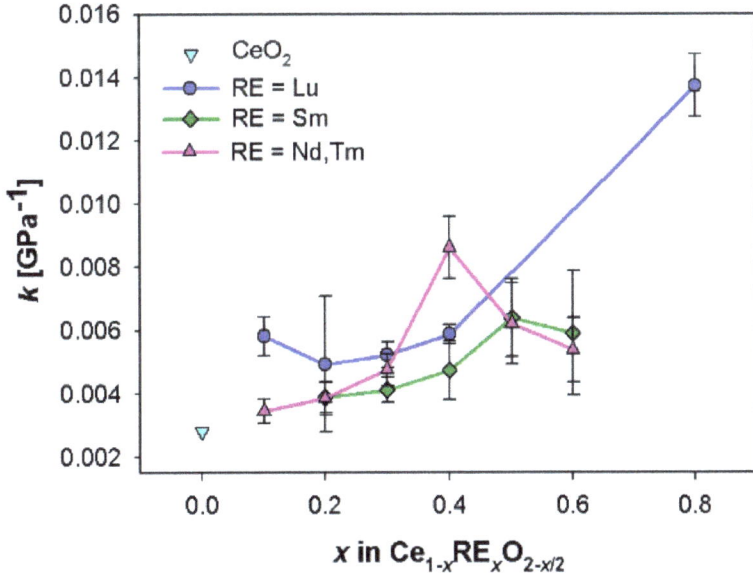

Figure 7. Compressibility k as a function of the doping ion amount x for each studied system.

5. Conclusions

This work critically reviews papers recently appeared in the literature dealing with the use of in situ high pressure X-ray diffraction in the design of doped ceria thin films to be mainly used as solid electrolytes in solid oxide electrochemical microdevices; it discusses the results therein reported highlighting the role of the bulk modulus of the bulk material in the choice of the most proper compositions of films.

The tensile strain possibly arising at the film/substrate interface gives a significant contribution to ionic conduction in doped ceria thin films, provided that it is not released through the occurrence of dislocations; compressibility is therefore a measure of the strain amount which can be tolerated by the structure. Within this scenario, the evaluation of bulk modulus through the refinement of the material equation of state, and hence of compressibility, is of help. Results obtained from Sm-, (Nd,Tm)- and Lu-doped ceria show that the smaller the doping ions, the higher compressibility, as a consequence of the higher binding energy of defect clusters containing smaller rare earths.

Supplementary Materials: The following are available online at https://www.mdpi.com/article/10.3390/coatings11060724/s1, Table S1: Hybrid structural model compared to the F model typical of CeO_2 and the C model typical of sesquioxides of heavy rare earths, such as Tm_2O_3.

Author Contributions: Conceptualization, C.A.; methodology, S.M. and C.A.; investigation, S.M.; resources, C.A.; writing—original draft preparation, S.M., A.C. and C.A.; writing—review and editing, S.M., A.C. and C.A.; project administration, C.A.; funding acquisition, C.A. All authors have read and agreed to the published version of the manuscript.

Funding: This research was funded by Compagnia di San Paolo, in the frame of the project COELUS—ID ROL: 32604.

Institutional Review Board Statement: Not applicable.

Informed Consent Statement: Not applicable.

Data Availability Statement: No new data were created or analyzed in this study. Data sharing is not applicable to this article.

Acknowledgments: Authors would like to thank the Elettra synchrotron facility for the provision of beamtime.

Conflicts of Interest: The authors declare no conflict of interest. The funders had no role in the design of the study; in the collection, analyses, or interpretation of data; in the writing of the manuscript, or in the decision to publish the results.

References

1. Mogensen, M.; Sammes, N.M.; Tompsett, G.A. Physical, chemical and electrochemical properties of pure and doped ceria. *Solid State Ion.* **2000**, *129*, 63–94. [CrossRef]
2. Steele, B.C.H. Appraisal of $Ce_{1-y}Gd_yO_{2-y/2}$ electrolytes for IT-SOFC operation at 500 °C. *Solid State Ion.* **2000**, *129*, 95–110. [CrossRef]
3. Spiridigliozzi, L.; Di Bartolomeo, E.; Dell'Agli, G.; Zurlo, F. GDC-based infiltrated electrodes for solid oxide electrolyzer cells (SOECs). *Appl. Sci.* **2020**, *10*, 3882. [CrossRef]
4. Beie, H.-J.; Gnörich, A. Oxygen gas sensors based on CeO_2 thick and thin films. *Sens. Actuators B Chem.* **1991**, *4*, 393–399. [CrossRef]
5. Gerblinger, J.; Lohwasser, W.; Lampe, U.; Meixner, H. High temperature oxygen sensor based on sputtered cerium oxide. *Sens. Actuators B Chem.* **1995**, *26*, 93–96. [CrossRef]
6. Yang, J.J.; Strukov, D.B.; Stewart, D.R. Memristive devices for computing. *Nat. Nanotechnol.* **2013**, *8*, 13–24. [CrossRef] [PubMed]
7. Kant, K.M.; Esposito, V.; Pryds, N. Strain induced ionic conductivity enhancement in epitaxial $Ce_{0.9}Gd_{0.1}O_{2-\delta}$ thin films. *Appl. Phys. Lett.* **2012**, *100*, 033105. [CrossRef]
8. Burbano, M.; Marrocchelli, D.; Watson, G.W. Strain effects on the ionic conductivity of Y-doped ceria: A simulation study. *J. Electroceramics* **2014**, *32*, 28–36. [CrossRef]
9. Rupp, J.L.; Infortuna, A.; Gauckler, L.J. Thermodynamic stability of gadolinia-doped ceria thin film electrolytes for micro-solid oxide fuel cells. *J. Am. Ceram. Soc.* **2007**, *90*, 1792–1797. [CrossRef]
10. Rainwater, B.H.; Velisavljevic, N.; Park, C.; Sun, H.; Waller, G.H.; Tsoi, G.M.; Vohra, Y.K.; Liu, M. High pressure structural study of samarium doped CeO_2 oxygen vacancy conductor—Insight into the dopant concentration relationship to the strain effect in thin film ionic conductors. *Solid State Ion.* **2016**, *292*, 59–65. [CrossRef]
11. Artini, C.; Joseph, B.; Costa, G.A.; Pani, M. Crystallographic properties of the Ce1-xLuxO2-x/2 system at pressures up to 7 GPa. *Solid State Ion.* **2018**, *320*, 152–158. [CrossRef]
12. Artini, C.; Massardo, S.; Carnasciali, M.M.; Joseph, B.; Pani, M. In situ high pressure structural investigation of sm-doped ceria. *Energies* **2020**, *13*, 1558. [CrossRef]
13. Artini, C.; Massardo, S.; Carnasciali, M.M.; Joseph, B.; Pani, M. Evaluation of the defect cluster content in singly and doubly doped ceria through in situ high-pressure X-ray diffraction. *Inorg. Chem.* **2021**, *60*, 7306–7314. [CrossRef]
14. Artini, C. RE-doped ceria systems and their performance as solid electrolytes: A puzzling tangle of structural issues at the average and local scale. *Inorg. Chem.* **2018**, *57*, 13047–13062. [CrossRef]
15. Inaba, H.; Tagawa, H. Ceria-based solid electrolytes. *Solid State Ion.* **1996**, *83*, 1–16. [CrossRef]
16. Kumar, S.A.; Kuppusami, P.; Yen-Pei, F. Structural, morphological and electrical properties of Sm-Gd Co-doped ceria thin films for micro-solid oxide fuel cells. *Mater. Lett.* **2020**, *275*, 128110. [CrossRef]
17. Bastakys, L.; Kalyk, F.; Marcinauskas, L.; Čyvienė, J.; Abakevičienė, B. Structural investigation of gadolinia-ceria multilayered thin films deposited by reactive magnetron sputtering. *Mater. Lett.* **2020**, *271*, 127762. [CrossRef]
18. Sanna, C.; Zhang, W.; Costamagna, P.; Holtappels, P. Synthesis and electrochemical characterization of $La_{0.6}Sr_{0.4}Co_{0.2}Fe_{0.8}O_{3-\delta}$/$Ce_{0.9}Gd_{0.1}O_{1.95}$ co-electrospun nanofiber cathodes for intermediate-temperature solid oxide fuel cells. *Int. J. Hydrog. Energy* **2021**, *46*, 13818–13831. [CrossRef]
19. Brito, M.E.; Morishita, H.; Yamada, J.; Nishino, H.; Uchida, H. Further improvement in performances of $La_{0.6}Sr_{0.4}Co_{0.2}Fe_{0.8}O_{3-\delta}$—doped ceria composite oxygen electrodes with infiltrated doped ceria nanoparticles for reversible solid oxide cells. *J. Power Sour.* **2019**, *427*, 293–298. [CrossRef]
20. Jiang, S.P. Development of lanthanum strontium cobalt ferrite perovskite electrodes of solid oxide fuel cells—A review. *Int. J. Hydrog. Energy* **2019**, *44*, 7448–7493. [CrossRef]
21. Jiang, S.P. Development of lanthanum strontium manganite perovskite cathode materials of solid oxide fuel cells: A review. *J. Mater. Sci.* **2008**, *43*, 6799–6833. [CrossRef]
22. Marcucci, A.; Zurlo, F.; Sora, I.N.; Placidi, E.; Casciardi, S.; Licoccia, S.; Di Bartolomeo, E. A redox stable Pd-doped perovskite for SOFC applications. *J. Mater. Chem. A* **2019**, *7*, 5344–5352. [CrossRef]
23. Kumar, K.L.A.; Durgajanani, S.; Jeyaprakash, B.G.; Rayappan, J.B.B. nanostructured ceria thin film for ethanol and triethylamine sensing. *Sens. Actuators B Chem.* **2013**, *177*, 19–26. [CrossRef]

24. Gupta, S.; Kuchibhatla, S.; Engelhard, M.; Shutthanandan, V.; Nachimuthu, P.; Jiang, W.; Saraf, L.; Thevuthasan, S.; Prasad, S. Influence of samaria doping on the resistance of ceria thin films and its implications to the planar oxygen sensing devices. *Sens. Actuators B Chem.* **2009**, *139*, 380–386. [CrossRef]
25. Ueda, T.; Defferriere, T.; Hyodo, T.; Shimizu, Y.; Tuller, H.L. Nanostructured Pr-doped Ceria (PCO) thin films as sensing electrodes in solid-electrolyte type gas sensors with enhanced toluene sensitivity. *Sens. Actuators B Chem.* **2020**, *317*, 128037. [CrossRef]
26. Liu, L.; Zang, C.; Wang, B.; Su, W.; Xiao, H.; Zhang, D.; Zhang, Y. Ceria thin film memristive device by magnetron sputtering method. *Vacuum* **2020**, *173*, 109128. [CrossRef]
27. Schmitt, R.; Spring, J.; Korobko, R.; Rupp, J.L. Design of oxygen vacancy configuration for memristive systems. *Acs Nano* **2017**, *11*, 8881–8891. [CrossRef]
28. Schweiger, S.; Pfenninger, R.; Bowman, W.J.; Aschauer, U.; Rupp, J.L.M. Designing strained interface heterostructures for memristive devices. *Adv. Mater.* **2017**, *29*, 1605049. [CrossRef]
29. Eyring, L. The binary rare earth oxides. In *Handbook on the Physics and Chemistry of Rare Earths*; Gschneidner, K.A., Jr., Eyring, L., Eds.; North Holland: Amsterdam, The Netherlands, 1979; Volume 3, pp. 337–399.
30. Costa, G.A.; Artini, C.; Ubaldini, A.; Carnasciali, M.M.; Mele, P.; Masini, R. Phase stability study of the pseudobinary system Gd_2O_3-Nd_2O_3 (T≤1350 °C). *J. Therm. Anal. Calorim.* **2008**, *92*, 101–104. [CrossRef]
31. Artini, C.; Costa, G.A.; Pani, M.; Lausi, A.; Plaisier, J. Structural characterization of the CeO_2/Gd_2O_3 mixed system by synchrotron X-ray diffraction. *J. Solid State Chem.* **2012**, *190*, 24–28. [CrossRef]
32. Artini, C.; Presto, S.; Massardo, S.; Pani, M.; Carnasciali, M.M.; Viviani, M. Transport properties and high temperature raman features of heavily Gd-doped ceria. *Energies* **2019**, *12*, 4148. [CrossRef]
33. Artini, C.; Presto, S.; Viviani, M.; Massardo, S.; Carnasciali, M.M.; Gigli, L.; Pani, M. The role of defects association in structural and transport properties of the $Ce_{1-x}(Nd_{0.74}Tm_{0.26})_xO_{2-x/2}$ system. *J. Energy Chem.* **2021**, *60*, 494–502. [CrossRef]
34. Omar, S.; Wachsman, E.D.; Jones, J.L.; Nino, J.C. Crystal structure-ionic conductivity relationships in doped ceria systems. *J. Am. Ceram. Soc.* **2009**, *92*, 2674–2681. [CrossRef]
35. Gupta, M.; Shirbhate, S.; Ojha, P.; Acharya, S. Processing and conductivity behavior of La, Sm, Fe singly and doubly doped ceria: As electrolytes for IT-SOFCs. *Solid State Ion.* **2018**, *320*, 199–209. [CrossRef]
36. Minervini, L.; Zacate, M.O.; Grimes, R.W. Defect cluster formation in M_2O_3-doped CeO_2. *Solid State Ion.* **1999**, *116*, 339–349. [CrossRef]
37. Li, Z.-P.; Mori, T.; Zou, J.; Drennan, J. Defects clustering and ordering in di- and trivalently doped ceria. *Mater. Res. Bull.* **2013**, *48*, 807–812. [CrossRef]
38. Presto, S.; Artini, C.; Pani, M.; Carnasciali, M.M.; Massardo, S.; Viviani, M. Ionic conductivity and local structural features in $Ce_{1-x}Sm_xO_{2-x/2}$. *Phys. Chem. Chem. Phys.* **2018**, *20*, 28338–28345. [CrossRef] [PubMed]
39. Koettgen, J.; Dück, G.; Martin, M. The oxygen ion conductivity of Lu doped ceria. *J. Phys. Condens. Matter* **2020**, *32*, 265402. [CrossRef]
40. Artini, C.; Pani, M.; Lausi, A.; Masini, R.; Costa, G.A. High temperature structural study of Gd-doped ceria by synchrotron X-ray Diffraction (673 K ≤ T ≤ 1073 K). *Inorg. Chem.* **2014**, *53*, 10140–10149. [CrossRef]
41. Artini, C.; Pani, M.; Carnasciali, M.M.; Buscaglia, M.T.; Plaisier, J.R.; Costa, G.A. Structural features of Sm- and Gd-doped ceria studied by synchrotron X-ray diffraction and μ-Raman spectroscopy. *Inorg. Chem.* **2015**, *54*, 4126–4137. [CrossRef]
42. Artini, C.; Pani, M.; Carnasciali, M.M.; Plaisier, J.R.; Costa, G.A. Lu-, Sm-, and Gd-doped ceria: A comparative approach to their structural properties. *Inorg. Chem.* **2016**, *55*, 10567–10579. [CrossRef] [PubMed]
43. Guan, X.; Zhou, H.; Liu, Z.; Wang, Y.; Zhang, J. High performance Gd^{3+} and Y^{3+} co-doped ceria-based electrolytes for intermediate temperature solid oxide fuel cells. *Mater. Res. Bull.* **2008**, *43*, 1046–1054. [CrossRef]
44. Coles-Aldridge, A.V.; Baker, R. Ionic conductivity in multiply substituted ceria-based electrolytes. *Solid State Ion.* **2018**, *316*, 9–19. [CrossRef]
45. Omar, S.; Wachsman, E.D.; Nino, J.C. Higher conductivity Sm^{3+} and Nd^{3+} co-doped ceria-based electrolyte materials. *Solid State Ion.* **2008**, *178*, 1890–1897. [CrossRef]
46. Arabaci, A.; Altınçekiç, T.G.; Der, M.; Öksüzömer, M.A.F. Preparation and properties of ceramic electrolytes in the Nd and Gd co-doped ceria systems prepared by polyol method. *J. Alloys Compd.* **2019**, *792*, 1141–1149. [CrossRef]
47. Venkataramana, K.; Madhuri, C.; Madhusudan, C.; Reddy, Y.; Bhikshamaiah, G.; Reddy, C. Investigation on La^{3+} and Dy^{3+} co-doped ceria ceramics with an optimized average atomic number of dopants for electrolytes in IT-SOFCs. *Ceram. Int.* **2018**, *44*, 6300–6310. [CrossRef]
48. Venkataramana, K.; Madhuri, C.; Madhusudan, C.; Reddy, Y.S.; Reddy, C.V. Investigation on micro-structural, structural, electrical and thermal properties of La^{3+}, Sm^{3+} & Gd^{3+} triple-doped ceria as solid-electrolyte for intermediate temperature-solid oxide fuel cell applications. *J. Appl. Phys.* **2019**, *126*, 144901. [CrossRef]
49. Rai, A.; Mehta, P.; Omar, S. Ionic conduction behavior in $Sm_xNd_{0.15-x}Ce_{0.85}O_{2-\delta}$. *Solid State Ion.* **2014**, *263*, 190–196. [CrossRef]
50. Spiridigliozzi, L.; Dell'Agli, G.; Marocco, A.; Accardo, G.; Pansini, M.; Yoon, S.P.; Ham, H.C.; Frattini, D. Engineered co-eprecipitation chemistry with ammonium carbonate for scalable synthesis and sintering of improved $Sm_{0.2}Ce_{0.8}O_{1.90}$ and $Gd_{0.16}Pr_{0.04}Ce_{0.8}O_{1.90}$ electrolytes for IT-SOFCs. *J. Ind. Eng. Chem.* **2018**, *59*, 17–27. [CrossRef]
51. Harrington, G.F.; Kim, S.; Sasaki, K.; Tuller, H.L.; Grieshammer, S. Strain-modified ionic conductivity in rare-earth substituted ceria: Effects of migration direction, barriers, and defect-interactions. *J. Mater. Chem. A* **2021**, *9*, 8630–8643. [CrossRef]

52. De Souza, R.A.; Ramadan, A.; Hörner, S. Modifying the barriers for oxygen-vacancy migration in fluorite-structured CeO_2 electrolytes through strain: A computer simulation study. *Energy Environ. Sci.* **2011**, *5*, 5445–5453. [CrossRef]
53. Shi, Y.; Garbayo, I.; Muralt, P.; Rupp, J.L.M. Micro-solid state energy conversion membranes: Influence of doping and strain on oxygen ion transport and near order for electrolytes. *J. Mater. Chem. A* **2017**, *5*, 3900–3908. [CrossRef]
54. Schweiger, S.; Kubicek, M.; Messerschmitt, F.; Murer, C.; Rupp, J.L.M. A microdot multilayer oxide device: Let us tune the strain-ionic transport interaction. *ACS Nano* **2014**, *8*, 5032–5048. [CrossRef] [PubMed]
55. Sun, L.; Marrocchelli, D.; Yildiz, B. Edge dislocation slows down oxide ion diffusion in doped CeO_2 by segregation of charged defects. *Nat. Commun.* **2015**, *6*, 6294. [CrossRef] [PubMed]
56. Sanna, S.; Esposito, V.; Pergolesi, D.; Orsini, A.; Tebano, A.; Licoccia, S.; Balestrino, G.; Traversa, E. Fabrication and electrochemical properties of epitaxial samarium-doped ceria films on $SrTiO_3$-buffered MgO substrates. *Adv. Funct. Mater.* **2009**, *19*, 1713–1719. [CrossRef]
57. Korte, C.; Keppner, J.; Peters, A.; Schichtel, N.; Aydin, H.; Janek, J. Coherency strain and its effect on ionic conductivity and diffusion in solid electrolytes—An improved model for nanocrystalline thin films and a review of experimental data. *Phys. Chem. Chem. Phys.* **2014**, *16*, 24575–24591. [CrossRef] [PubMed]
58. Garcia-Barriocanal, J.; Rivera-Calzada, A.; Varela, M.; Sefrioui, Z.; Iborra, E.; Leon, C.; Pennycook, S.J.; Santamaria, J. Colossal ionic conductivity at interfaces of epitaxial ZrO_2:Y_2O_3/$SrTiO_3$ heterostructures. *Science* **2008**, *321*, 676–680. [CrossRef]
59. Kilner, J.A. Feel the strain. *Nat. Mater.* **2008**, *7*, 838–839. [CrossRef]
60. Chen, L.; Chen, C.L.; Chen, X.; Donner, W.; Liu, S.W.; Lin, Y.; Huang, D.X.; Jacobson, A.J. Electrical properties of a highly oriented, textured thin film of the ionic conductor Gd:$CeO_{2-\delta}$ on (001) MgO. *Appl. Phys. Lett.* **2003**, *83*, 4737–4739. [CrossRef]
61. Sanna, S.; Esposito, V.; Tebano, A.; Licoccia, S.; Traversa, E.; Balestrino, G. Enhancement of ionic conductivity in Sm-doped ceria/yttria-stabilized zirconia heteroepitaxial structures. *Small* **2010**, *6*, 1863–1867. [CrossRef]
62. Jiang, J.; Shen, W.; Hertz, J. Structure and ionic conductivity of nanoscale gadolinia-doped ceria thin films. *Solid State Ion.* **2013**, *249*, 139–143. [CrossRef]
63. Christie, G.M.; van Berkel, F.P.F. Microstructure—ionic conductivity relationships in ceria-gadolinia electrolytes. *Solid State Ion.* **1996**, *83*, 17–27. [CrossRef]
64. Suzuki, T.; Kosacki, I.; Anderson, H.U. Microstructure—electrical conductivity relationships in nanocrystalline ceria thin films. *Solid State Ion.* **2002**, *151*, 111–121. [CrossRef]
65. Rupp, J.L.M.; Gauckler, L.J. Microstructures and electrical conductivity of nanocrystalline ceria-based thin films. *Solid State Ion.* **2006**, *177*, 2513–2518. [CrossRef]
66. Karageorgakis, N.I.; Heel, A.; Rupp, J.L.M.; Aguirre, M.H.; Graule, T.; Gauckler, L.J. Properties of flame sprayed $Ce_{0.8}Gd_{0.2}O_{1.9-\delta}$ electrolyte thin films. *Adv. Funct. Mater.* **2010**, *21*, 532–539. [CrossRef]
67. Joo, J.H.; Choi, G.M. Electrical conductivity of thin film ceria grown by pulsed laser deposition. *J. Eur. Ceram. Soc.* **2007**, *27*, 4273–4277. [CrossRef]
68. Chiodelli, G.; Malavasi, L.; Massarotti, V.; Mustarelli, P.; Quartarone, E. Synthesis and characterization of $Ce_{0.8}Gd_{0.2}O_{2-y}$ polycrystalline and thin film materials. *Solid State Ion.* **2005**, *176*, 1505–1512. [CrossRef]
69. Schlupp, M.V.F.; Kurlov, A.; Hwang, J.; Yáng, Z.; Döbeli, M.; Martynczuk, J.; Prestat, M.; Son, J.W.; Gauckler, L.J. Gadolinia doped ceria thin films prepared by aerosol assisted chemical vapor deposition and applications in intermediate-temperature solid oxide fuel cells. *Fuel Cells* **2013**, *5*, 658–665. [CrossRef]
70. Lair, V.; Živković, L.S.; Lupan, O.; Ringuedé, A. Synthesis and characterizathion of electrodeposited samaria and samaria-doped ceria thin films. *Electrochim. Acta* **2011**, *56*, 4638–4644. [CrossRef]
71. Lee, Y.H.; Chang, I.; Cho, G.Y.; Park, J.; Yu, W.; Tanveer, W.H.; Cha, S.W. Thin film solid oxide fuel cells operating below 600 C: A review. *Int. J. Precis. Eng. Manuf.* **2018**, *5*, 441–453. [CrossRef]
72. Artini, C.; Cingolani, A.; Anselmi Tamburini, U.; Valenza, F.; Latronico, G.; Mele, P. Effect of the sintering pressure on structure and microstructure of the filled skutterudite $Sm_y(Fe_xNi_{1-x})_4Sb_{12}$. *Mater. Res. Bull.* **2021**, *139*, 111261. [CrossRef]
73. Ghadiri, M.; Kang, A.K.; Gorji, N.E. XRD characterization of graphene-contacted perovskite solar cells: Moisture degradation and dark-resting recovery. *Superlattices Microstruct.* **2020**, *146*, 106677. [CrossRef]
74. Shen, Y.; Kumar, R.S.; Pravica, M.; Nicol, M.F. Characteristics of silicone fluid as a pressure transmitting medium in diamond anvil cells. *Rev. Sci. Instrum.* **2004**, *75*, 4450–4454. [CrossRef]
75. Artini, C.; Carnasciali, M.; Viviani, M.; Presto, S.; Plaisier, J.; Costa, G.; Pani, M. Structural properties of Sm-doped ceria electrolytes at the fuel cell operating temperatures. *Solid State Ion.* **2018**, *315*, 85–91. [CrossRef]
76. Ubaldini, A.; Artini, C.; Costa, G.A.; Carnasciali, M.M.; Masini, R. Synthesis and thermal decomposition of mixed Gd-Nd oxalates. *J. Therm. Anal. Calorim.* **2008**, *91*, 797–803. [CrossRef]
77. Hammersley, A.P. FIT2D: An Introduction and Overview. *ESRF Internal Report*; ESRF: Grenoble, France, 1997; ESRF97HA02T.
78. Rodríguez-Carvajal, J. Recent advances in magnetic structure determination by neutron powder diffraction. *Phys. B Condens. Matter* **1993**, *192*, 55–69. [CrossRef]
79. Rietveld, H.M. A profile refinement method for nuclear and magnetic structures. *J. Appl. Cryst.* **1969**, *2*, 65–71. [CrossRef]
80. Shannon, R.D. Revised effective ionic radii and systematic studies of interatomic distances in halides and chalcogenides. *Acta Crystallogr.* **1976**, *32*, 751–767. [CrossRef]

81. Artini, C.; Carnasciali, M.M.; Plaisier, J.R.; Costa, G.A.; Pani, M. A novel method for the evaluation of the Rare Earth (RE) coordination number in RE-doped ceria through Raman spectroscopy. *Solid State Ion.* **2017**, *311*, 90–97. [CrossRef]
82. Murnaghan, F.D. The compressibility of media under extreme pressures. *Proc. Natl. Acad. Sci. USA* **1944**, *30*, 244–247. [CrossRef]
83. Birch, F. Finite elastic strain of cubic crystals. *Phys. Rev.* **1947**, *71*, 809–824. [CrossRef]
84. Poirier, J.-P.; Tarantola, A. A logarithmic equation of state. *Phys. Earth Planet. Inter.* **1998**, *109*, 1–8. [CrossRef]
85. Vinet, P.; Smith, J.R.; Ferrante, J.; Rose, J.H. Temperature effects on the universal equation of state of solids. *Phys. Rev. B* **1987**, *35*, 1945–1953. [CrossRef]
86. Gonzalez-Platas, J.; Alvaro, M.; Nestola, F.; Angel, R. EosFit7-GUI: A new graphical user interface for equation of state calculations, analyses and teaching. *J. Appl. Cryst.* **2016**, *49*, 1377–1382. [CrossRef]
87. Anderson, O.L.; Nafe, J.E. The bulk modulus-volume relationship for oxides compounds and related geophysical problems. *J. Geophys. Res.* **1965**, *70*, 3951–3963. [CrossRef]
88. Hill, S.; Catlow, C. A hartree-fock periodic study of bulk ceria. *J. Phys. Chem. Solids* **1993**, *54*, 411–419. [CrossRef]

Article

Synthesis and Properties of p-Si/n-Cd$_{1-x}$Ag$_x$O Heterostructure for Transparent Photodiode Devices

Mannarsamy Anitha [1], Karuppiah Deva Arun Kumar [2], Paolo Mele [2,*], Nagarajan Anitha [3], Karunamoorthy Saravanakumar [4], Mahmoud Ahmed Sayed [5,6], Atif Mossad Ali [5,6] and Lourdusamy Amalraj [7]

[1] Department of Physics, Sri Vidhya College of Arts and Science, Virudhunagar 626005, India; m.anithakrishnan.1986@gmail.com
[2] Shibaura Institute of Technology, College of Engineering, Saitama 337-8570, Japan; apj.deva1990@gmail.com
[3] Department of Physics, Devanga Arts College, Arupukottai 626101, India; toanithaganesh@gmail.com
[4] Department of Chemistry, VHNSN College, Virudhunagar Tamilnadu 626001, India; sravanan205@gmail.com
[5] Department of Physics, Faculty of Science, King Khalid University, Abha 61413, Saudi Arabia; frrag75@gmail.com (M.A.S.); atifali@kku.edu.sa (A.M.A.)
[6] Physics Department, Faculty of Science, Al-Azher University, Assiut 71524, Egypt
[7] Department of Physics, VHNSN College, Virudhunagar 626001, India; amalrajprofessor@gmail.com
* Correspondence: pmele@shibaura-it.ac.jp

Abstract: We developed silver-doped Cd$_{1-x}$Ag$_x$O thin films (where x = 0, 0.01, 0.02, 0.03 and 0.04) on amorphous glass substrate by an automated nebulizer spray pyrolysis set-up. The XRD patterns show rock salt cubic crystal structures, and the crystallite sizes vary with respect to Ag doping concentrations. SEM images exhibited a uniform distribution of grains with the addition of Ag; this feature could support the enhancement of electron mobility. The transmittance spectra reveal that all films show high transmittance in the visible region with the observed bandgap of about 2.40 eV. The room temperature photoluminescence (PL) studies show the increase of near-band-edge (NBE) emission of the films prepared by different Ag doping levels, resulting in respective decreases in the bandgaps. The photodiode performance was analyzed for the fabricated p-Si/n-Cd$_{1-x}$Ag$_x$O devices. The responsivity, external quantum efficiency and detectivity of the prepared p-Si/n-Cd$_{1-x}$Ag$_x$O device were investigated. The repeatability of the optimum (3 at.% Ag) photodiode was also studied. The present investigation suggests that Cd$_{1-x}$Ag$_x$O thin films are the potential candidates for various industrial and photodetector applications.

Keywords: Cd$_{1-x}$Ag$_x$O thin films; p-Si/n-Cd$_{1-x}$Ag$_x$O; PL; photodiode

1. Introduction

Thin film technology is a quick-moving field which includes the application of structures, film processing and the manufacture of devices by controlling the shape and size of grains at the nanometer range [1]. Nanometer-range thin films have distinctive and specific physical properties associated with their size [2]. Metal oxide semiconductors have attracted increasing attention because of their wide range of applications in various devices, which include optical communications, photodiode, gas sensors, low emissive windows and solar cells [3]. In the last decade, researchers developed metal oxide-based transparent detectors such as gas sensors and photodetectors [4]. At present, photodetectors are widely studied for use in imaging scanners, switching, touch panels and light sensors. Zheng et al. [4] studied the flexible ZnO–CdO nanofiber arrays for UV photodetectors with 95% transparency. Metal oxides are very common and easily available on Earth, and they are used to fabricate transparent devices due to their low absorption in the visible region. The typical wide bandgap metal oxides such as SnO$_2$, ZnO, TiO$_2$ and In$_2$O$_3$ can perform as ultraviolet light sensors [5]. The visible range-absorbing CdO is commonly used to fabricate solar cell devices, and it also might be a potential candidate to detect the light spectrum [5].

CdO is an n-type semiconductor with a direct optical bandgap of 2.3–2.5 eV [6] and nonstoichiometric composition, owing to the defects of oxygen vacancies or cadmium interstitials that act as double-loaded donors [7,8]. The unique qualities of CdO metal oxide are its low cost, chemical stability and wide availability.

In general, different kinds of nanostructures offer more attractive properties due to their high surface-to-volume ratios, consequently enhancing the light-sensing ability. The different nanostructures can be grown by changing several parameters, such as the chemical synthesis process, rate of the growth time, substrate temperature, nature of the substrate, and doping. However, metal doping can effectively modify the nature of grains in the host lattice; consequently, it could affect the surface-to-volume ratio or the observing capability of light on the host film surface, whereas the doping can enhance the grain growth level and Debye length, resulting in improved optoelectrical properties. Moreover, doping is a key parameter to improve the performance of any semiconducting materials and used in wide range of device applications [9]. Previously, CdO systems were prepared with different doping elements such as aluminum [10], manganese [11], gallium [12], indium [13], zinc [14] and silver [15] for various applications. However, the silver (Ag) metal element has been commonly used because of its numerous good characteristics and its notable chemical reactivity in solutions [16]. Ag is one of the active metals that might be easily oxidized by the doping process. Therefore, it is very interesting to understand the role of silver in a transparent CdO matrix. Recently, scientists worked on pure and Ag-doped CdO nanostructures through various techniques [15,17–19]. To our knowledge, the silver-doped CdO thin film prepared by nebulizer spray pyrolysis has not been reported. Kumar et al. [20,21] reported the advantages of the nebulizer spray pyrolysis (NSP) technique. The main advantages of the NSP technique are good adherence, mechanical stability, being inexpensive, fine droplet formation and capability for mass production for large area deposition that are required for industrial device applications.

Therefore, the aim of the present study is to fabricate $Cd_{1-x}Ag_xO$ thin films (where x = 0, 0.01, 0.02, 0.03 and 0.04) via an automated nebulizer spray pyrolysis method. Ag doping induced the improvement of the optical and electrical properties for the performance of solar cells and other optical devices. Recently, Kathalingam et al. [22] reported the fabrication of an n-Sn:CdO/p-Si heterostructure for visible photodetectors. They suggested that a high-bandgap n-type semiconductor can be easily coated on p-Si substrates to fabricate heterojunction solar cells. In addition, the fabricated p-Si/n-CdO heterojunction could decrease the recombination loss due to a mixture of small-bandgap absorber material and high-bandgap window material. Based on the above work, we report the development of the n-Ag:CdO/p-Si heterostructure and study the photoresponse. This work is propounded for the first time and suggests that the CdO thin films could be utilized for the development of multilayer photodetectors.

2. Materials and Methods

Ag-doped CdO films were prepared on glass substrates using the automated nebulizer spray pyrolysis technique. A total of 0.1 M of high-purity (3N) cadmium acetate dehydrate $(Cd(CH_3(COO)_2)\cdot 2H_2O)$ was liquefied by using a mixture of isopropyl alcohol and deionized water, with a volume ratio of 1:3 (20 mL). Silver nitrate $(AgNO_3)$ was used as a dopant precursor with different concentrations, as given in Supplementary Table S1. The prepared precursor solution was sprayed onto the heated glass substrates with a surface area of 25 mm^2. The heated substrates were maintained at a constant temperature of 200 °C +/− 2 °C using a PID (select 500) controller. The parameters were maintained during spray deposition, including a carrier gas (compressed air) pressure of 14.7×10^4 Nm^{-2}, a distance between the nozzle and substrate of 10 mm and a solution flow rate of 0.5 mL/min, which were previously optimized. When the sprayed solution fell on the heated substrates, thermal decomposition took place on the film surface, resulting in yellow-colored CdO particles being grown. The structural properties of films were analyzed by an X'PERT-PRO diffractometer (Malvern Pananalytical Ltf, Malvern, UK) (λ = 0.15406 nm) with CuKα radiation. The SEM investigation was performed by a scanning electron microscope (TESCAN VEGA-III, Tescan, Brno, Czech Republic). The elemental analysis was carried out using energy dispersive X-ray spectroscopy (EDX). The optical properties of the films were characterized by a UV-visible spectrophotometer (Shimadzu-UV 1800 model, Shimadzu, Kyoto, Japan). The reflectance spectrum of the samples was recorded by a UV-2400 PC Series. The photoluminescence (PL) spectra was recorded using a fluorescence spectrophotometer (HITACHI F-7000), Hitachi, Tokyo, Japan with an excitation wavelength of ~325 nm. A 532 nm laser wavelength was used as illumination light to measure the photosensitivity, external quantum efficiency (EQE) and detectivity values with the help of a Keithley source meter (model 2450, Tectronix Inc, Beaverton, OR, USA).

3. Results and Discussion
3.1. Structural Analysis

The phase structure of the Ag-doped $Cd_{1-x}Ag_xO$ films (where x = 0, 0.01, 0.02, 0.03 and 0.04) was studied by XRD, which is shown in Figure 1. The identified diffraction peaks $2\theta_{(111)} = 33.03°$, $2\theta_{(200)} = 38.44°$, $2\theta_{(220)} = 55.34°$, $2\theta_{(311)} = 66.03°$ and $2\theta_{(222)} = 69.35°$ matched well with the cubic structure of CdO (space group: Fm3m) (JCPDS Card No. 05-0640).

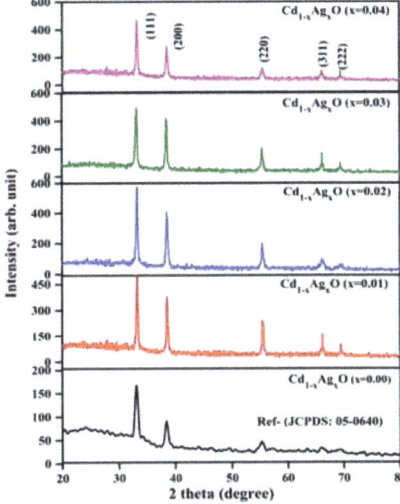

Figure 1. XRD patterns of Ag-doped CdO thin films with different Ag doping levels.

In Figure 1, a highly intense peak obtained at $2\theta = 33.03°$ for all CdO films indicates a preferential growth orientation along the (111) direction. After the inclusion of Ag, the intensity of the XRD characteristic peaks increased gradually up to the optimum doping level (Ag = 0.03), due to Cd^{2+} ions being replaced by Ag^+ ions because of the larger ionic radius of Ag^+ (0.126 nm) compared with Cd^{2+} (0.095 nm). The increase in peak intensity or decrease of the full width at half maximum (β) indicated the enhancement of film crystallinity and grain sizes [23]. The comparative cure for 2 at.% and 3 at.% Ag-doped CdO films are given in Figure S1. The figure confirms that the (111) plane increased with the increasing Ag doping level. However, by increasing the Ag doping beyond the optimum level (Ag = 0.03), the peak intensity and film crystallinity started to decrease (Ag = 0.04). This may be attributed to the higher solution quantity or Ag precipitation, resulting a powder-like film being formed on the film surface. In addition, the peak reduction may have possibly occurred by the slight variation of both the Cd and Ag ionic radii, which led to the creation of lattice distortion or crystal imperfections [24]. It is worth noting that there were no impurity phase formations related to Ag or Ag_3O_4, even for a higher dopant level up to 4%. Rex et al. [24] reported the same trend in Ag-doped PbS (4% Ag) films under the NSP method. This could justify that the doping of Ag effectively substituted the site of Cd, resulting the formation of single-phase CdO. As expected, there was no observable peak shift in the predominant (111) plane, confirming the structural purity of CdO.

The lattice parameters were calculated using the standard relation of the cubic structure [25]. The calculated values increased with respect to the increase in Ag concentration up to 3% Ag and then reduced for higher doping levels, and they are tabulated in Table 1.

Table 1. Structural properties of $Cd_{1-x}Ag_xO$ thin films.

Ag Doping Level (at.%)	Crystallite Size (nm)	Strain ε	Lattice Constant a = b = c (Å)	Unit Cell Volume (v) (Å)^3	TC
0	28	0.0042	4.6828	102.69	1.76
1	29	0.0040	4.6842	102.78	1.93
2	31	0.0038	4.6854	102.86	2.18
3	34	0.0034	4.6869	102.97	2.35
4	32	0.0038	4.6893	103.11	1.97

This behavior could be due to the change in the concentration of the native defects or structural defects developed in the prepared host films. The structural parameters, such as the crystallite size (D) and strain (ε), were determined using standard relations [26,27]. The calculated D and ε values of the $Cd_{1-x}Ag_xO$ films with respect to different Ag concentrations are given in Table 1. The crystallite size enhanced with the increase in Ag content, and then it slightly reduced for higher Ag (4 at.%) doping levels. This could possibly vary with the change of structural defects like ε. The defect factors of ε decreased initially as a function of the increasing Ag up to 3% and then slightly increased for the 4% Ag-doped CdO thin film. A decrease in defect values could be attributed to the enhancement of crystallinity caused by the periodic arrangement of atoms in the host crystal structure.

The texture coefficient TC $_{(hkl)}$ values demonstrate the crystal orientation of the (111) plane of the prepared $Cd_{1-x}Ag_xO$ system by using X-ray data with the equation given below [26]:

$$TC_{(hkl)} = \frac{I_{(hkl)}/I_{0(hkl)}}{N_r^{-1} \sum I_{(hkl)}/I_{0(hkl)}} \qquad (1)$$

where $I_{(hkl)}$, $I_{0(hkl)}$ and N denote the usual meanings. From Table 1, it is found that the calculated TC values increased with the increasing Ag doping concentration and then finally decreased. The obtained TCs for all the films were higher than 1 (TC > 1), which indicates the preferable grain growth in certain orientations. This result suggests that a lot of crystallites were gathered along the (111) plane, which helped improve the electrical performance. The variation of TC values was the same as in our previous report [26]. The XRD studies showed that the Ag, up to 3 at.%, maintained good structural quality (crystallinity) with minimum defects (strain) compared with that of other prepared films.

3.2. Surface Morphology Analysis

The observed SEM images of the pure CdO and $Cd_{1-x}Ag_xO$ films are shown in Figure S2. The pure CdO film showed densely packed, brain-like grains [28] uniformly distributed on the entire film surface, whereas the bunch of irregular spherical grains without the presence of voids were observed in 1–2 at.% Ag-doped films. The film with 3 at.% Ag doping exhibited spherical shaped grains with well-defined grain boundaries [29]. The particles were further agglomerated by increasing the doping beyond the optimum level (4 at.% Ag). Moreover, the grain sizes increased by increasing the doping concentrations up to 3 at.%. Thereafter, the grain size slightly decreased for higher-level doping (4 at.% Ag). Thus, various levels of Ag doping concentrations effectively modified the host CdO film surface, and this observation could be correlated with the XRD results. The change in grain size signified the change of grain boundaries in the host lattice. The composition of the pure and Ag-doped CdO (3 at.%) films were confirmed by energy dispersive X-ray spectroscopy (EDX). The obtained EDX spectra of the prepared films are shown in the inset of Figure S2. The expected elements of Cd, O and Ag are present with the desired atomic percentages which were taken. The composition percentages of Cd, O, and Ag were 48.07%, 49.06%, and 2.87%, respectively, for 3 at.% Ag.

3.3. Optical Analysis

The optical properties were explored with a UV-visible spectrophotometer. The optical transmittance and reflectance spectra of $Cd_{1-x}Ag_xO$ films are given in Figure 2a,b, respectively. From Figure 2a, the obtained transparency was ~70% in the visible wavelength range. Transmittance was reduced slightly with the increasing doping concentration, which may have been due to the decrement in oxygen vacancies. Furthermore, the lessening of transmittance was due to the change in grain size and film thickness [30]. The observed transmittance range confirmed that the prepared film could work as a transparent photodetector.

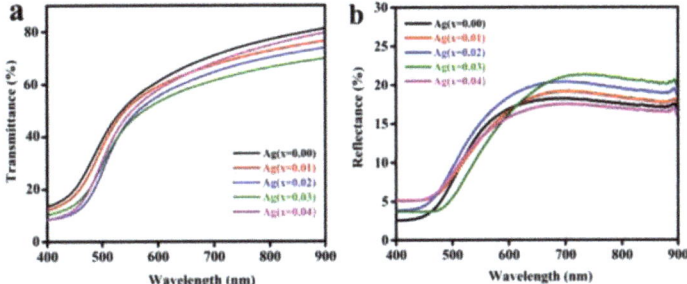

Figure 2. (a) Transmittance and (b) reflectance spectra of pure and Ag-doped CdO thin films.

The transmittance value was not as close to the lower wavelength of the visible region due to the appearance of absorption. Furthermore, the primary absorption edge was shifted in the direction of the higher wavelength (redshift) with respect to the Ag concentration. This edge in the wavelength can decide the variation of the bandgap because the transmittance is associated with stimulating the charge carriers from the valence to conduction bands. The optical bandgap is related by the following equation [24]:

$$\alpha h\nu = B(h\nu - E_g)^n \qquad (2)$$

where B is a constant and E_g is the bandgap of the material. The direct bandgaps of the $Cd_{1-x}Ag_xO$ films were obtained from a plot of $(\alpha h\nu)^2$ versus $h\nu$, as illustrated in Figure 3. The bandgap values of the prepared films are given in Table 2. The E_g value of the pure CdO was 2.45 eV, which matched well with the one reported by Usharani et al. [31], whereas the E_g values for the doped CdO films were found to decrease from 2.43 eV to 2.37 eV with the increase in the Ag doping level. This might be due to the substitution of Ag in the sites of Cd. A similar reduction in the bandgap was observed for the Ag-doped PbS film under the NSP method [24]. Moreover, redshift absorption (toward a higher wavelength) further confirmed the decrease of the bandgap. For a higher doping concentration (4 at.% Ag), the bandgap was slightly enhanced due to the doped Ag ions effectively modifying the band structure of the host CdO lattice. The reduction of film crystallinity (XRD) and blueshift of the near-band-edge (NBE) emission (PL) occurred from 3 at.% to 4 at.% Ag-doped CdO, as evidenced by the increase of the bandgap. As shown in Table 2, the bandgap was low for the film doped with 3 at.% Ag, which absorbed more visible light than the other films. Consequently, the photoelectric interaction was enhanced, resulting in improving the photosensor performance with respect to the applied voltage.

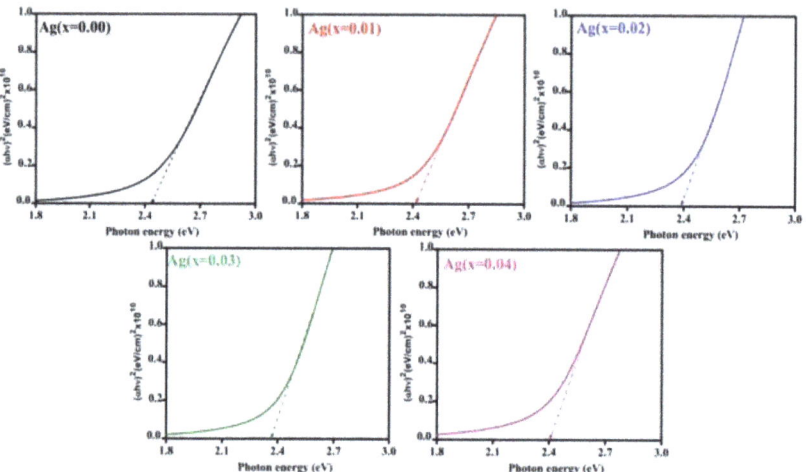

Figure 3. Optical bandgap spectra of pure and Ag-doped CdO thin films.

Table 2. Optical properties of $Cd_{1-x}Ag_xO$ thin films.

Ag Doping Level (at.%)	For λ = Average of 600–900 nm			Bandgap E_g (eV)
	T%	R%	n	
0	75	17	2.41	2.45
1	71	19	2.53	2.43
2	68	20	2.58	2.40
3	65	22	2.62	2.37
4	70	16	2.36	2.41

Other optical factors, such as the refractive index (n) and reflectance (R), play an important role for any optical device. The R and n values of the $Cd_{1-x}Ag_xO$ thin films are listed in Table 2. The R values of the Ag-doped CdO thin films increased from 17% to 22% as a function of the doping concentration increasing from 1 at.% to 3 at.% Ag. The variation of reflectance was in oppositional behavior to the transmittance with respect to the doping concentration, as shown in Figure 2b. The refractive index n was determined from the values of reflectance and the extinction coefficient (k) using the following equation [32]:.

$$n = \frac{(1+R)}{(1-R)} + \sqrt{\frac{4R}{(1-R)^2} - k^2} \quad (3)$$

From Table 2, the observed low value of n confirmed the smooth surface morphologies and minimum reflectance of the prepared films. The variation of n systematically increased with the increasing Ag doping level due to the decrease of transmittance (T). Commonly, the n variation had the opposite trend of T, which may be attributed to the increase of light scattering of the photons. The estimated n values were in the range of 2.36–2.62 in the visible range for different doping concentration, as shown in Table 2.

3.4. Photoluminescence Analysis

Photoluminescence (PL) was used to investigate the factor of the electron hole pairs in the semiconductor materials [33].

Figure 4 presents the photoluminescence spectra of the prepared $Cd_{1-x}Ag_xO$ films obtained by an excitation wavelength of ~325 nm. The room temperature PL spectra mainly consisted of one strong green emission band observed at ~520 nm (2.38 eV) in all the prepared films. The green emission was consistent with the near-band-edge emission (NBE), which originated from the recombination of the electron and holes. In addition, the observed NBE emission by PL correlated well with the bandgap obtained by the UV-visible study. Similar green emission was previously observed by Velusamy et al. [23] for CdO films. The overall peak intensity increased as a function of the Ag doping concentration. This might be representative of the increasing quantity of Ag ions on the host CdO lattice and the decrease in space between the Ag^+ and Cd^{2+} ions, while for the higher doping level, the peak intensity decreased slightly, which may have been due to the increasing interstitial placement level of the Ag ions near the Cd sites. The NBE emission shifted gradually toward longer wavelength sides (from 517 to 522 nm) with increasing Ag doping levels. Consequently, there was a decrease in the energy gap, as evidenced from Figure 3. Gaussian curve fitting was used to find the additional defect peaks from the prepared $Cd_{1-x}Ag_xO$ films. From the fitted PL cure, a peak was observed at 521 nm for all the prepared films. This might have been the presence of Cd as interstitials or native state defects. As seen from the figure, the defect peaks were considerably increased with respect to the Ag doping concentration, and as a result, it could help to enhance the photoelectric performance. In addition, the other low-intensity green emission peak was identified at ~545 nm for all the films, which arose due to the presence of defects like oxygen vacancies, as is clearly shown in Figure 4. The film with 3 at.% Ag doping had more oxygen vacancy at the film surface than the others. Therefore, it could react well with the photon from the laser light. Moreover, the Ag doping induced a redshift, which may have been due to the surface area [34], resulting in a reduction of the bandgap. We believe that the Ag dopant can enhance the surface area and crystalline quality. The above factors might be evidence for enhancing light absorption and improving charge carrier generation. A similar result was observed for Nd-doped ZnO thin films by Paul et al. [35].

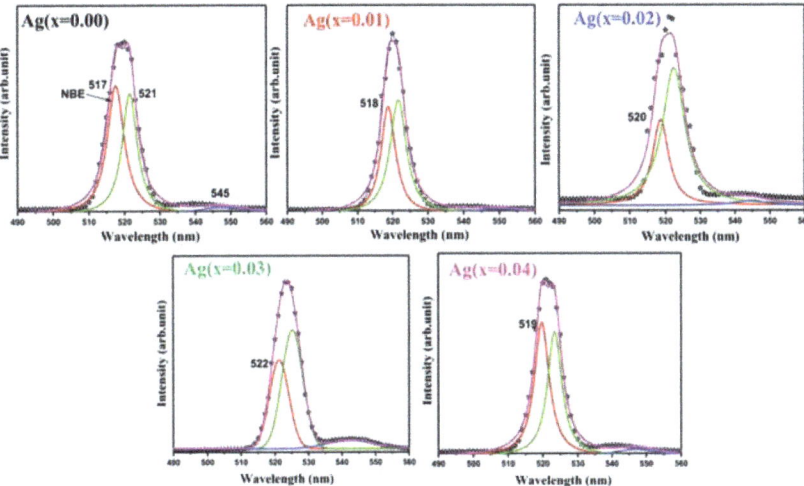

Figure 4. Gaussian curve-fitted photoluminescence spectra of pure and Ag-doped CdO thin films.

3.5. Photodiode Preparation and Properties

The schematic diagram of the fabricated heterojunction photodiode p-Si/n-$Cd_{1-x}Ag_xO$/Ag structure is shown in Figure 5. Before fabricating the heterostructure device, the native oxide should be removed from the Si substrate. Here, we used $HF:H_2O$ (1:10) solution treatment and then rinsed it with deionized water. The Ag-doped CdO films were coated on the *p*-type silica substrate by the NSP method to form a heterojunction device. Silver paste was used as contact electrodes on both the *p*-Si and *n*-CdO surfaces to measure the photocurrent with respect to the bias voltage.

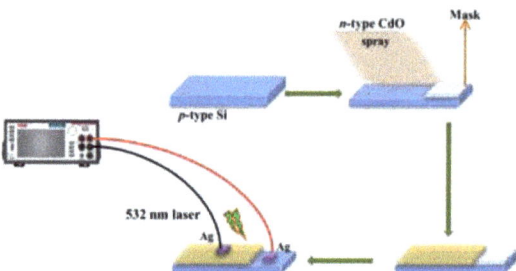

Figure 5. The schematic diagram of the prepared *p*-Si/*n*-CdO:Ag structure.

The photocurrent performance of the prepared photodiodes was measured using a ~532 nm laser with a Keithley source meter (model-2450) at room temperature. The current–voltage (I–V) characteristics were investigated at different bias voltages from -5 V to $+5$ V under dark and UV illumination conditions. The measured I–V spectra of the *p*-Si/*n*-Cd$_{1-x}$Ag$_x$O/Ag structure are given in Figure 6. The observed dark and light currents both increased gradually with respect to the increase in applied voltage. However, the current level was higher for light illumination compared to the dark condition, due to the increase of electrons when the diode interacted with the incident photons obtained from light. The photocurrent (under light) was enhanced significantly for both the forward and reverse biases, indicating successful light sensing for the fabricated photodiodes. It was observed that the photocurrent was nearly two orders of magnitude higher than the dark current, likely due to the inner photoelectric effect (i.e., when a photon of sufficient energy falls on the photodiode, it creates an electron hole pair).

The external quantum efficiency (EQE), responsivity (R*) and specific detectivity (D*) values were calculated using the standard relations, which were $EQE = \frac{Rhc}{e\lambda}$, $R* = \frac{I_p}{A \times P_{in}}$ and $D* = R\sqrt{\frac{A}{2eI_d}}$ [35,36]. The calculated D* and EQE values were directly proportional to R, which increased for the Ag-doped CdO photodiodes, as shown in Figure 7.

The prepared photodiode responsivity increased from 0.05 AW^{-1} to 0.28 AW^{-1} for the pure CdO to 3 at.% Ag-doped CdO photodiode and then decreased to 0.03 AW^{-1} for the higher doping of Ag (4 at.%). The maximum external quantum efficiency was ~98.21%, observed for the optimum CdO photodiode (3 at.% Ag-doped CdO) when compared with others. Ravikumar et al. [37] obtained an EQE value of 73.03% for a 5% Nd-doped CdO, which was a bit lower than our value. The specific detectivity was enhanced gradually from 5.07×10^9 to 1.10×10^{11} (jones) for the pure to 3 at.% Ag-doped CdO photodiode. This observed specific detectivity was comparably higher than the previously reported one for CdS:Ag/Si [38]. The observed good photoelectric properties may have been due to maximum electron hole pair generation when it interacted with high-energy photons.

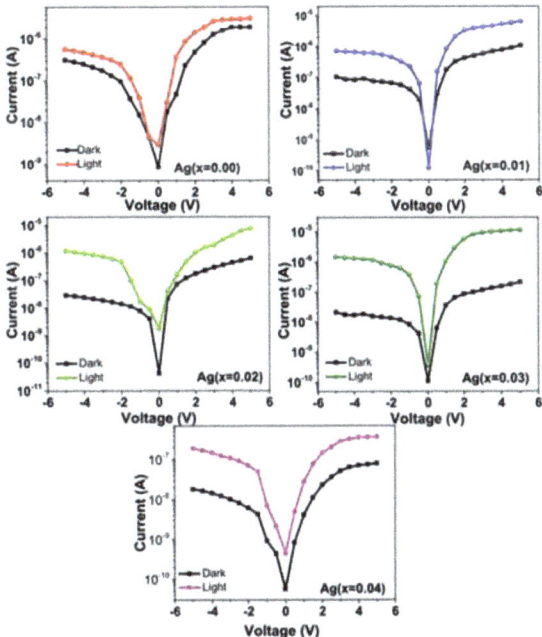

Figure 6. Light and dark current–voltage (I–V) characteristics of the fabricated the p-Si/n-Cd$_{1-x}$Ag$_x$O/Ag structures.

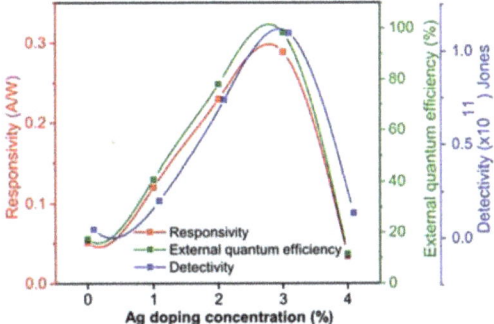

Figure 7. The variations of the responsivity (R), specific detectivity (D*) and external quantum efficiency (EQE) with respect to different Ag doping amounts.

The linear dynamic range (LDR) is one of the important factors to recognize the performance of any photodiode and photodetectors, and it can be estimated using the following equation [39]:

$$LDR = 20 \log \left(\frac{I_{min}}{I_{max}} \right) \quad (4)$$

where I_{min} and I_{max} represent the dark current and light current, respectively. The observed LDR of the prepared p-Si/n-Cd$_{1-x}$Ag$_x$O/Ag structure was 36.7 dB for the 3 at.% Ag-doped CdO. This value was close to the ZnO-based p-n junction diode [39], though it is worth noting that the CdO-based visible photodiode prepared by a simple spray method showed a reasonable LDR value compared with that of the other *p-n* junction [39,40]. Hence, we believe that this is an essential move to fabricate the flexible photodiodes.

To find the ideality factor, the *I–V* characteristics of the optimum *p*-Si/*n*-CdO:Ag photodiode (3 at.% Ag) under dark and light conditions are given in Figure 8. The observed photocurrent exponentially increased when compared with the dark current. In this work, the observed photocurrent level (11.6 × 10^{-6} A at 5V) was higher than the previously reported values. Recently, Xiao et al. [41] fabricated the Ga_2O_3:CdO-based amorphous thin-film transistor and observed a photocurrent of about 20 nA at 20 V.

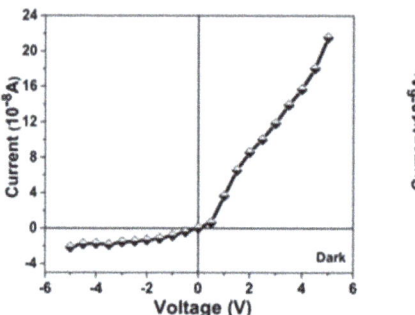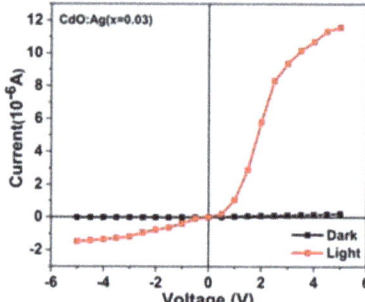

Figure 8. *I-V* characteristics of the fabricated *p*-Si/*n*-CdO:Ag photodiode measured in dark and illuminated conditions at a bias voltage between −5 and +5 V.

The ideality factor (η) is one of the important parameters for any heterostructure diode for finding the recombination mechanisms ruling inside the diode. Based on the theory of thermionic emission (TE), the ideality factor (η) can be calculated with the use of dark *I–V* curves by the following relation [42]:

$$I = I_0 \exp\left(\frac{eV}{\eta kT} - 1\right) \quad (5)$$

where *e* is the electron charge, *V* is the voltage, I_0 is the saturation current, *k* is the Boltzmann constant and *T* is the absolute temperature (300 K). The η value of the diode was calculated from the slope of the linear part at the forward bias curves, and the value was 3.2. For an ideal diode, the η is equal to one, which means that the current flow is due to thermionic emission, though the reported value is commonly higher than unity. In our case, the ideality factor value was found to be higher than unity, indicating the existence of interface states or surface defects on the native oxide layer. This can possibly happen from thermionic emission and other electron transport mechanisms. Tunneling recombination in particular could play a leading role in controlling the junction behavior and increasing the η [43]. Yakuphanoglu et al. [44] observed an ideality factor of 5.41 for a nanocluster-based n-CdO/*p*-Si photodiode.

In order to understand the photosensing performance of the optimal photodiode (3 at.% Ag-doped CdO), the photocurrent experiment was performed under different light intensities from 1 mW/cm² to 5 mW/cm² at 0.5 V, as shown in Figure 9a. When the diode interacted with different intensities of light (turning on the light), the current quickly reached a specific level, and when the light was turned off, the photocurrent reached its beginning stage. This was because of the increase of the number of photo-generated charge carriers and the free electrons that supported the increase in the current for the light-on condition, whereas the numbers of free electrons went down when the light was off [45]. The photocurrent varied between the light-on and off conditions due to the trapping of deep levels of charge carriers [46]. The figure shows that the current values were increasing gradually with the increasing intensities of light from 1 mW/cm² to 5 mW/cm². The time taken for increasing and decreasing the current level was measured from the fabricated p-Si/n-CdO:Ag structure. The increase in time was considered to be ~90% of the time taken to reach the maximum value, and the reverse was the decrease in time. The increase in time (τ_{inc}) and decrease in time (τ_{dec}) were 0.8 s and 1.0 s, respectively.

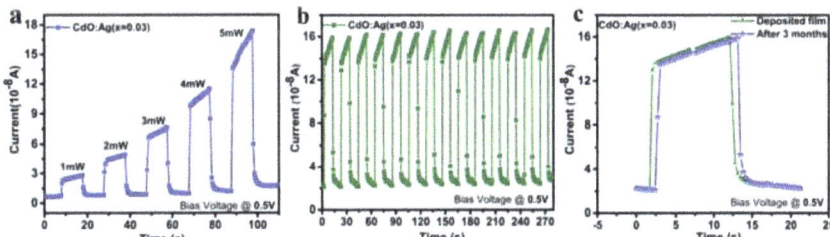

Figure 9. (**a**) Photocurrent kinetics with respect to different light intensities from 1 mW/cm^2 to 5 mW/cm^2. (**b**) Repeatability and (**c**) stability at 5 mW/cm^2 for the optimum 3 at.% Ag-doped CdO photodiode.

To understand the photodetection capability (repeatability) of the prepared heterostructure, a constant laser power (5 mW/cm^2) was used as a function of the increasing time. Figure 9b shows the transient photocurrent response of the prepared heterostructure under a bias voltage of 0.5 V. The transient photosensing performances maintained the same range with respect to the increase in time, which confirmed good sensing capability and repeatability. Figure 9c illustrates the variation of the current for fresh and 3 months-aged p-Si/n-CdO:Ag photodiodes to define the stability of the optimum sensor. It was found that the electrical current range reached above 95% of the fresh photodiode at 0.5 V, while the response time was a bit slower (1.2 s) when compared with the fresh photodiode (0.8 s). At the same time, the recovery time was nearly equal (~1.0 s) for both the fresh and aged photodiodes. Hence, we conclude that the aged photodiode still showed good photosensing performance and good stability (reproducibility) even after 3 months.

4. Conclusions

In summary, the successful preparation of transparent $Cd_{1-x}Ag_xO$ thin films on glass and p-Si substrates for photodiode investigation. The XRD pattern revealed the formation of a cubic structured CdO lattice with a (111) preferential orientation, and there were no peak shifts in all the films, which confirmed the absence of an impurity phase. The SEM images revealed that all the films had spherical grains, and the grain size increased with the increase in the Ag doping level. EDX analysis revealed that the Cd, O and Ag elements were present in the prepared films. The undoped film showed the maximum transmittance of ~75%, and the corresponding bandgap was 2.45 eV, which confirmed the transparent nature of the CdO material. The obtained bandgap decreased with an increase in the doping level. From the present study, it is evident that the fabricated p-Si/n-CdO:Ag photodiode showed an improved photoresponse in the visible region.

Supplementary Materials: Attached The following are available online at https://www.mdpi.com/article/10.3390/coatings11040425/s1, Figure S1: The comparative cure for 2 at.% and 3 at.% of Ag doped CdO thin films, Figure S2. The scanning electron micrographs for the prepared CdO thin films with EDX spectrum for pure and 3 at.% Ag. Table S1: Preparation of Ag doping concentrations for CdO thin films (0.1 M).

Author Contributions: Conceptualization, methodology, data analysis and writing—original draft, M.A.; project supervision, writing and approval for submission of the manuscript, L.A.; interpretation of data, writing—review and editing, K.D.A.K. and P.M.; interpretation of data—review and editing, N.A. and K.S.; discussion and reviewing of the manuscript, M.A.S. and A.M.A. All authors have read and agreed to the published version of the manuscript.

Funding: The authors from KKU extended their appreciation to the Deanship of Scientific Research at King Khalid University for funding this work through the research group program under grant number R.G.P.1/298/42.

Institutional Review Board Statement: Not applicable.

Informed Consent Statement: Not applicable.

Data Availability Statement: Data are contained within the article.

Conflicts of Interest: The authors declare no conflict of interest.

References

1. Lopez-Serrano, A.; Muñoz-Olivas, R.; Sanz-Landaluze, J.; Olasagasti, M.; Rainieri, S.; Camara, C. Comparison of bioconcentration of ionic silver and silver nanoparticles in zebrafish eleutheroembryos. *Environ. Pollut.* **2014**, *191*, 207–214. [CrossRef] [PubMed]
2. Auffan, M.; Rose, J.; Bottero, J.-Y.; Lowry, G.V.; Jolivet, J.-P.; Wiesner, M.R. Towards a definition of isnorganic nanoparticles from an environmental, health and safety perspective. *Nat. Nanotechnol.* **2009**, *4*, 634–641. [CrossRef] [PubMed]
3. Sivakumar, S.; Venkatesan, A.; Soundhirarajan, P.; Khatiwada, C.P. Synthesis, characterizations and anti-bacterial activities of pure and Ag doped CdO nanoparticles by chemical precipitation method. *Spectrochim. Acta Part A Mol. Biomol. Spectrosc.* **2015**, *136*, 1751–1759. [CrossRef]
4. Zheng, Z.; Gan, L.; Li, H.; Ma, Y.; Bando, Y.; Golberg, D.; Zhai, T. A Fully Transparent and Flexible Ultraviolet–Visible Photodetector Based on Controlled Electrospun ZnO-CdO Heterojunction Nanofiber Arrays. *Adv. Funct. Mater* **2015**, *25*, 5885–5894. [CrossRef]
5. Rajput, J.K.; Pathak, T.K.; Swart, H.C.; Purohit, L.P. Synthesis of CdO Nanoflowers by Sol-Gel Method on Different Substrates with Photodetection Application. *Phys. Status Solidi* **2019**, *216*, 1900093. [CrossRef]
6. Chopra, S.R. *Thin Film Solar Cells*; Plenum Press: New York, NY, USA, 1993.
7. Minami, T.; Ida, S.; Miyata, T. High rate deposition of transparent conducting oxide thin films by vacuum arc plasma evaporation. *Thin Solid Film.* **2002**, *416*, 92–96. [CrossRef]
8. Pan, L.L.; Li, G.Y.; Lian, J.S. Structural, optical and electrical properties of cerium and gadolinium doped CdO thin films. *Appl. Surf. Sci.* **2013**, *274*, 365–370. [CrossRef]
9. Aydın, C.; El-Nasser, H.M.; Yakuphanoglu, F.; Yahia, I.S.; Aksoy, M. Nanopowder synthesis of aluminum doped cadmium oxide via sol–gel calcination processing. *J. Alloy Compd.* **2011**, *509*, 854–858. [CrossRef]
10. Wongcharoen, N.; Gaewdang, T.; Wongcharoen, T. Electrical Properties of Al-Doped CdO Thin Films Prepared by Thermal Evaporation in Vacuum. *Energy Procedia* **2012**, *15*, 361–370. [CrossRef]
11. Gao, W.; Yang, S.; Yang, S.; Lv, L.; Du, Y. Synthesis and magnetic properties of Mn doped CuO nanowires. *Phys. Lett. A* **2010**, *375*, 180–182. [CrossRef]
12. Thambidurai, M.; Muthukumarasamy, N.; Ranjitha, A.; Velauthapillai, D. Structural and optical properties of Ga-doped CdO nanocrystalline thin films. *Superlattices Microstruct.* **2015**, *86*, 559–563. [CrossRef]
13. Zheng, B.; Hu, W. Influence of substrate temperature on the structural and properties of In-doped CdO films prepared by PLD. *J. Semicond.* **2013**, *34*, 053003. [CrossRef]
14. Ravichandran, A.T.; Robert Xavier, A.; Pushpanathan, K.; Nagabhushana, B.M.; Chandramohan, R. Structural and optical properties of Zn doped CdO nanoparticles synthesized by chemical precipitation method. *J. Mater. Sci. Mater. Electron.* **2016**, *27*, 2693–2700. [CrossRef]
15. Yüksel, M.; Şahin, B.; Bayansal, F. Nano structured CdO films grown by the SILAR method: Influence of silver-doping on the morphological, structural and optical properties. *Ceram. Int.* **2016**, *42*, 6010–6014. [CrossRef]
16. Malachová, K.; Praus, P.; Rybková, Z.; Kozák, O. Antibacterial and antifungal activities of silver, copper and zinc montmorillonites. *Appl. Clay Sci.* **2011**, *53*, 642–645. [CrossRef]
17. Salem, A. Silver-doped cadmium oxide nanoparticles: Synthesis, structural and optical properties. *Eur. Phys. J. Plus* **2014**, *129*, 263. [CrossRef]
18. Majid, A.; Afzal, Z.; Murtaza, S.; Nabi, G.; Ahmad, N. Synthesis and Characterization of Silver Doped Cadmium Oxide Nanoparticles. *J. Adv. Phys.* **2013**, *2*, 116–118. [CrossRef]
19. Saravanakumar, K.; Muthuraj, V.; Jeyaraj, M. The design of novel visible light driven Ag/CdO as smart nanocomposite for photodegradation of different dye contaminants. *Spectrochim. Acta Part A Mol. Biomol. Spectrosc.* **2018**, *188*, 291–300. [CrossRef]
20. Deva Arun Kumar, K.; Mele, P.; Ponraj, J.S.; Haunsbhavi, K.; Varadharajaperumal, S.; Alagarasan, D.; Algarni, H.; Angadi, B.; Murahari, P.; Ramesh, K. Methanol solvent effect on photosensing performance of AZO thin films grown by nebulizer spray pyrolysis. *Semicond. Sci. Technol.* **2020**, *35*, 085013. [CrossRef]
21. Deva Arun Kumar, K.; Valanarasu, S.; Tamilnayagam, V.; Amalraj, L. Structural, morphological and optical properties of SnS2 thin films by nebulized spray pyrolysis technique. *J. Mater. Sci. Mater. Electron.* **2017**, *28*, 14209–14216. [CrossRef]
22. Kathalingam, A.; Kesavan, K.; Rana, A.U.H.S.; Jeon, J.; Kim, H.-S. Analysis of Sn Concentration Effect on Morphological, Optical, Electrical and Photonic Properties of Spray-Coated Sn-Doped CdO Thin Films. *Coatings* **2018**, *8*, 167. [CrossRef]

23. Velusamy, P.; Babu, R.R.; Ramamurthi, K.; Dahlem, M.S.; Elangovan, E. Highly transparent conducting cerium incorporated CdO thin films deposited by a spray pyrolytic technique. *RSC Adv.* **2015**, *5*, 102741–102749. [CrossRef]
24. Rosario, S.R.; Kulandaisamy, I.; Kumar, K.D.A.; Ramesh, K.; Ibrahium, H.A.; Awwad, N.S. Ag-doped PbS thin films by nebulizer spray pyrolysis for solar cells. *Int. J. Energy Res.* **2020**, *44*, 4505–4515. [CrossRef]
25. Velusamy, P.; Babu, R.R.; Ramamurthi, K.; Elangovan, E.; Viegas, J.; Dahlem, M.S.; Arivanandhan, M. Characterization of spray pyrolytically deposited high mobility praseodymium doped CdO thin films. *Ceram. Int.* **2016**, *42*, 12675–12685. [CrossRef]
26. Anitha, M.; Saravanakumar, K.; Anitha, N.; Amalraj, L. Influence of a novel co-doping (Zn + F) on the physical properties of nano structured (1 1 1) oriented CdO thin films applicable for window layer of solar cell. *Appl. Surf. Sci.* **2018**, *443*, 55–67. [CrossRef]
27. Haunsbhavi, K.; Deva Arun Kumar, K.; Mele, P.; Aldossary, O.M.; Ubaidullah, M.; Mahesh, H.M.; Murahari, P.; Angadi, B. Pseudo n-type behaviour of nickel oxide thin film at room temperature towards ammonia sensing. *Ceram. Int.* **2021**. [CrossRef]
28. Bagheri Khatibani, A.; Hallaj, Z.A.; Rozati, S.M. Some physical properties of CdO:F thin films prepared by spray pyrolysis. *Eur. Phys. J. Plus* **2015**, *130*, 254. [CrossRef]
29. Thirumoorthi, M.; Prakash, J.T.J. A study of Tin doping effects on physical properties of CdO thin films prepared by sol–gel spin coating method. *J. Asian Ceram. Soc.* **2016**, *4*, 39–45. [CrossRef]
30. Dakhel, A.A. Correlated transport and optical phenomena in Ga-doped CdO films. *Sol. Energy* **2008**, *82*, 513–519. [CrossRef]
31. Usharani, K.; Balu, A.R.; Nagarethinam, V.S.; Suganya, M. Characteristic analysis on the physical properties of nanostructured Mg-doped CdO thin films—Doping concentration effect. *Prog. Nat. Sci. Mater. Int.* **2015**, *25*, 251–257. [CrossRef]
32. Anand, V.; Sakthivelu, A.; Karuppiah, D.A.K.; Valanarasu, S.; Ganesh, V.; Shkir, M.; AlFaify, S.; Algarni, H. Rare earth Eu3+ co-doped AZO thin films prepared by nebulizer spray pyrolysis technique for optoelectronics. *J. Sol-Gel Sci. Technol.* **2018**, *86*, 293–304. [CrossRef]
33. Balachandran, S.; Praveen, S.G.; Velmurugan, R.; Swaminathan, M. Facile fabrication of highly efficient, reusable heterostructured Ag–ZnO–CdO and its twin applications of dye degradation under natural sunlight and self-cleaning. *RSC Adv.* **2014**, *4*, 4353–4362. [CrossRef]
34. Farhat, O.F.; Halim, M.M.; Ahmed, N.M.; Qaeed, M.A. ZnO nanofiber (NFs) growth from ZnO nanowires (NWs) by controlling growth temperature on flexible Teflon substrate by CBD technique for UV photodetector, Superlattice. *Microst* **2016**, *100*, 1120. [CrossRef]
35. Poul Raj, I.L.; Valanarasu, S.; Hariprasad, K.; Ponraj, J.S.; Chidhambaram, N.; Ganesh, V.; Ali, H.E.; Khairy, Y. Enhancement of optoelectronic parameters of Nd-doped ZnO nanowires for photodetector applications. *Opt. Mater.* **2020**, *109*, 110396. [CrossRef]
36. Shkir, M.; Ashraf, I.M.; Khan, A.; Khan, M.T.; El-Toni, A.M.; AlFaify, S. A facile spray pyrolysis fabrication of Sm:CdS thin films for high-performance photodetector applications. *Sens. Actuators A Phys.* **2020**, *306*, 111952. [CrossRef]
37. Ravikumar, M.; Chandramohan, R.; Kumar, K.D.A.; Valanarasu, S.; Ganesh, V.; Shkir, M.; Alfaify, S.; Kathalingam, A. Effect of Nd doping on structural and opto-electronic properties of CdO thin films fabricated by a perfume atomizer spray method. *Bull. Mater. Sci.* **2019**, *42*, 8. [CrossRef]
38. Najm, N.I.; Hassun, H.K.; al-Maiyaly, B.K.H.; Hussein, B.H.; Shaban, A.H. Highly selective CdS:Ag heterojunction for photodetector applications. *Aip Conf. Proc.* **2019**, *2123*, 020031. [CrossRef]
39. Hanna, B.; Surendran, K.P.; Narayanan Unni, K.N. Low temperature-processed ZnO thin films for p–n junction-based visible-blind ultraviolet photodetectors. *RSC Adv.* **2018**, *8*, 37365–37374. [CrossRef]
40. Gong, X.; Tong, M.; Xia, Y.; Cai, W.; Moon, J.S.; Cao, Y.; Yu, G.; Shieh, C.-L.; Nilsson, B.; Heeger, A.J. High-Detectivity Polymer Photodetectors with Spectral Response from 300 nm to 1450 nm. *Science* **2009**, *325*, 1665–1667. [CrossRef]
41. Xiao, X.; Liang, L.; Pei, Y.; Yu, J.; Duan, H.; Chang, T.-C.; Cao, H. Solution-processed amorphous Ga2O3:CdO TFT-type deep-UV photodetectors. *Appl. Phys. Lett* **2020**, *116*, 192102. [CrossRef]
42. Rhoderick, E.H.; Williams, R.H. *Metal-Semiconductor Contacts*; Clarendon Press Oxford: Oxford, UK, 1978.
43. Moun, M.; Kumar, M.; Garg, M.; Pathak, R.; Singh, R. Understanding of MoS(2)/GaN Heterojunction Diode and its Photodetection Properties. *Sci. Rep.* **2018**, *8*, 11799. [CrossRef] [PubMed]
44. Yakuphanoglu, F.; Caglar, M.; Caglar, Y.; Ilican, S. Electrical characterization of nanocluster n-CdO/p-Si heterojunction diode. *J. Alloy. Compd.* **2010**, *506*, 188–193. [CrossRef]
45. Gozeh, B.A.; Karabulut, A.; Yildiz, A.; Dere, A.; Arif, B.; Yakuphanoglu, F. SILAR Controlled CdS Nanoparticles Sensitized CdO Diode Based Photodetectors. *Silicon* **2020**, *12*, 1673–1681. [CrossRef]
46. Çiçek, O.; Tecimer, H.U.; Tan, S.O.; Tecimer, H.; Altındal, Ş.; Uslu, İ. Evaluation of electrical and photovoltaic behaviours as comparative of Au/n-GaAs (MS) diodes with and without pure and graphene (Gr)-doped polyvinyl alcohol (PVA) interfacial layer under dark and illuminated conditions. *Compos. Part B Eng.* **2016**, *98*, 260–268. [CrossRef]

Article

Construction of Rutile-TiO₂ Nanoarray Homojuction for Non-Contact Sensing of TATP under Natural Light

Yan Tang [1,†], Yuxiang Zhang [2,3,†], Guanshun Xie [1], Youxiong Zheng [1], Jianwei Yu [1], Li Gao [1,*] and Bingxin Liu [1,*]

1. Qinghai Provincial Key Laboratory of New Light Alloys, Qinghai Provincial Engineering Research Center of High Performance Light Metal Alloys and Forming, Qinghai University, Xining 810016, China
2. Key Laboratory of Materials Physics, Institute of Solid State Physics, Chinese Academy of Science, Hefei 230031, China
3. Department of Materials Science and Engineering, University of Science and Technology of China, Hefei 230026, China; zyxustc@mail.ustc.edu.cn
* Correspondence: 2007990030@qhu.edu.cn (L.G.); liubx408@nenu.edu.cn (B.L.); Fax: +86-9715310440 (B.L.)
† These authors contributed equally to this work.

Received: 23 March 2020; Accepted: 17 April 2020; Published: 20 April 2020

Abstract: Triacetone triperoxide (TATP) is a new terrorist explosive, and most nitrogen-based sensors fail to detect TATP. Herein, a sea urchin-like TiO_2-covered TiO_2 nanoarray is constructed as a TATP-sensitive homojunction (HJ) by one step hydrothermal method. By taking fluorine-doped tin oxide (FTO) and indium tin oxide (ITO) conducting glass as the substrate, the conducting glass is horizontally and vertically put in the reactor to epitaxially grow TiO_2–FTO, TiO_2–ITO, TiO_2–FTO–HJ and TiO_2–ITO–HJ. TiO_2–FTO–HJ shows a broad absorption band edge in the visible region and high sensitivity to TATP under the simulating natural light compared with TiO_2–FTO, TiO_2–ITO, and TiO_2–ITO–HJ. E-field intensity distribution simulation reveals that constructing homojunctions between the urchin-shaped TiO_2 nanosphere and TiO_2 nanoarrays can enhance the localized electromagnetic field intensity at the interface of junctions, which may provide photocatalysis active sites to reduce TATP molecules by promoting charge separation. Moreover, the TiO_2–FTO–HJ shows high selectivity to TATP among ammonium nitrate, urea and sulfur, which are common homemade explosive raw materials.

Keywords: TiO_2: homojunction; TATP; contactless detection

1. Introduction

In recent years, the bombing terrorist attack has become the main manifestation of terrorism activities. Public places with dense populations, such as railway stations, airports, subway stations, and so on, are becoming the first target of terrorists, and so carrying out explosive detection in public places is an important anti-terrorist measure. How to rapidly and accurately detect explosives in luggage parcels in real time has become an important topic of today's anti-terrorist activities in the international community [1–5].

Common explosives can be simply divided into standard explosives and non-standard explosives. Standard explosives mainly include 2,4,6-trinitrotoluene (TNT), 2,4-dinitrotoluene (DNT), p-nitrotoluene (PNT), picric acid (PA), etc. Non-standard explosives mainly include triacetone peroxide (TATP), Ammonium nitrate (AN), Urea, Sulfur, etc. There are various kinds of detection techniques for explosives at present [6–13]. Among these, dog sniffing method is one traditional detection method for tracing explosives, and its sensitivity is able to reach the parts per trillion (ppt) range. However, it has low efficiency. Police dogs generally need to rest for 15 to 30 min after continuously working for 30 min, and the training cost is high. In addition to the dog sniffing method, detection methods of explosives with more mature technology include the fluorescence detection method [14], electrochemical detection [15], ion spectroscopy [16], X-ray spectrometry, and so on [17]. However, these techniques also have disadvantages; for example, for ion spectroscopy and X-ray imaging technology, the equipment is heavy with high costs. Meanwhile, chromatographic analysis is limited to laboratory detection and cannot meet the requirements of field detection. In addition, most detection techniques cannot achieve the detection of non-standard explosives, and these disadvantages restrict the application in the actual detection of explosive. Moreover, several technologies have been discovered for the detection of trace amounts of TATP (Table S1).

Inspired by the dog sniffing method, a material used in gas-sensitive sensors for detecting explosives is produced by making use of the inherent properties of metal oxide semiconductors. In addition, gas-sensitive sensors are the most suitable for manufacturing portable explosive detection devices for explosives [17–19]. The detection of explosives is determined based on the electrical signal of the gas-sensitive sensor. One feasible way of improving the detection speed, sensitivity, accuracy and other attributes of the sensor is to magnify the contact area between the gas molecules and the sensor.

TiO_2 has good gas sensitivity and TiO_2 nanoarray can effectively enhance the contact area. Wang et al. [20] successfully prepared highly ordered aligned nitrogen-doped titania nanotube array films by liquid phase deposition (LPD) method, obtaining a new band gap corresponding to 2.17 eV in the visible region (570 Nm). Jie et al. [21] fabricated the silver ion-modified titanium nanotube array (AG/TiO_2-NT) by electrochemical anodization as an electrochemical energy storage electrode for supercapacitors. Jain et al. [22] developed sensors using gold nanoparticles (GnPs) and tubular TiO_2 for the detection of glycated hemoglobin in blood. The group of Hu [23] used anatase TiO_2 arrays with different crystal faces in different solvents by low-temperature solvothermal synthesis, which is mainly used in the field of photocatalysis. Cheng et al. [24] organized Pd nanoparticles on the surface of a carbon-coated TiO_2 nanowire array substrate using the potentiostatic pulse method, which is also used in the field of photocatalysis. Zhang et al. [25] developed TiO_2/ZnFe-LDH photoanode by photo-assisted electro deposition on TiO_2 nanoarray for the cracking performance of PEC water.

Although TiO_2 nanoarrays can be prepared by different methods, the above methods were not simple or effective in detecting gas sensing. In this work, the hydrothermal method is used to prepare a titanium dioxide nanoarray and to analyze the phase, morphology, ultraviolet absorption spectrum and gas-sensitive properties of the titanium dioxide nanoarray.

2. Materials and Methods

2.1. Materials

Acetone (99%, Sinopharm Ggroup Chemical Co. Ltd, Shanghai, China), ethanol (AR, Wokai, Shanghai, China), concentrated hydrochloric acid (AR, Wokai, Shanghai, China), titanium tetrachloride (99.5%, Oubokai, Shenzhen, China), ITO and FTO conductive glass (10 mm × 10 mm, Nozo, Luoyang, China) and deionized water were used.

2.2. Experimental

Preparation of TiO_2-FTO: FTO glass was subjected to ultrasonic cleaning in acetone, ethyl alcohol and deionized water for 15 min and dried. The FTO glass was vertically fixed on the Teflon-lined stainless autoclave wall. 10 mL of concentrated hydrochloric acid and 1 mL of titanium tetrachloride was continually stirred for 10 min. Then the above resulting mixture is transferred into a Teflon-lined stainless autoclave and maintained at 180 °C for 4 h. The system was then naturally cooled to ambient temperature. After that, the precipitation was sequentially washed with ethanol and distilled water to obtain TiO_2–FTO.

Preparation of TiO_2–FTO–HJ: The preparation process was similar to the process for TiO_2–FTO. The difference is that the FTO glass was horizontally fixed on the bottom of the Teflon-lined stainless autoclave.

The preparation process of TiO_2–ITO and TiO_2–FTO–HJ was similar to the process for TiO_2–FTO and TiO_2–FTO–HJ. The difference was that the FTO glass was replaced by ITO glass.

2.3. Characterization

The scanning electron microscope used in the experiment was JEOL's JSM-6106LV (Tokyo, Japan). The Bruker Corporation's X-ray diffractometer D8 Advance (Billerica, MA, USA) was used to analyze the sample, and Kα radial of the Cu target material was used (Wavelength λ = 1.5418 Å. voltage V = 40 kV and electricity I = 40 mA). Thermo Electron's Fourier transform infrared spectrum analyzer Nicolet 6700 (Waltham, MA, USA) was used to measure the infrared absorption spectrum. The measurement range was 4000–400 wavenumber (cm^{-1}). The UV-Vis absorption spectrum was determined using UV-4802S of Unico Instrument Co., Ltd. (Dayton, OH, USA), with a testing range of 300–800 nm. The Raman spectrum was recorded on Renishaw's microscopic confocal Raman spectrometer RM2000 (Gloucestershire, UK) with a laser wavelength of 514 nm.

The different analytes were evaporated in a 50 mL transparent chamber and the test was conducted at room-temperature in saturated ammonium nitrate (AN), urea and sulfur (S) vapor. Since the TATP has very high vapor pressure at room temperature, it is diluted by air to 600 ppb. The time-dependent photoresponse of the sensor film was determined in a conventional two electrode configuration and recorded by Keithley 4200A-SCS Parameter Analyzer (Portland, ME, USA) under the simulating natural light.

3. Results and Discussion

3.1. Morphology and Structural Characterization of TiO_2 Nanoarray

3.1.1. Morphology of TiO_2 Nanoarray

The FTO and ITO conducting glass is horizontally or vertically put into the reaction still in order to construct a TiO_2 nanoarray by a simple hydrothermal method. Figure 1 is the SEM images of TiO_2 nanoarray that puts FTO conducting glass into the reaction vessel vertically at 180 °C for 4 h. Figure 1a,c shows the front view and cross-section diagram of the nanoarray, while Figure 1b,d shows enlarged views of Figure 1a,c respectively. As can be seen from Figure 1a,b, the TiO_2 nanoarray is rod-shaped with uniform growth, and the length of the nanoarray is basically consistent. Combined with Figure 1c,d, the uniformity of the nanoarray can be clearly seen. The two sides of the nanoarray are basically parallel, and the thickness of the nanoarray is about 10.8 µm. The TiO_2 nanoarray grown on FTO conducting glass is recorded as TiO_2–FTO.

Figure 2 is the SEM images of the TiO_2 nanoarray that places the ITO conducting glass into the reaction vessel vertically at 180 °C for 4 h. Figure 2a,c shows the front view and cross-section diagram of the nanoarray, while Figure 2b,d shows enlarged views of Figure 2a,c, respectively. Compared to FTO (Figure 1a), the nanoarrays grown on the ITO as shown in Figure 2a are more loose. It can be seen from Figure 2b that there are obvious radial nanorods from the center of the urchin-shaped cluster, and the rod diameter and rod length are uniform. From Figure 2c,d, it can be seen more clearly from the cross-section that the thickness of the radial nanoarray is about 8.1 µm. The TiO_2 nanoarray grown on ITO conducting glass is recorded as TiO_2–ITO.

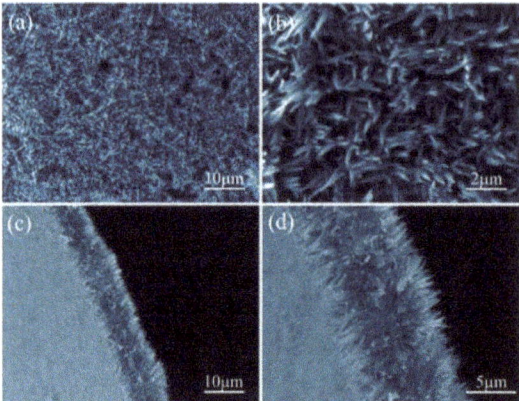

Figure 1. SEM images of TiO$_2$ nanoarrays grown on vertically placed FTO substrate: (**a**) Front view of TiO$_2$ nanoarray; (**b**) A magnified view; (**c**) TiO$_2$ nanoarray cross-section; (**d**) C magnified view.

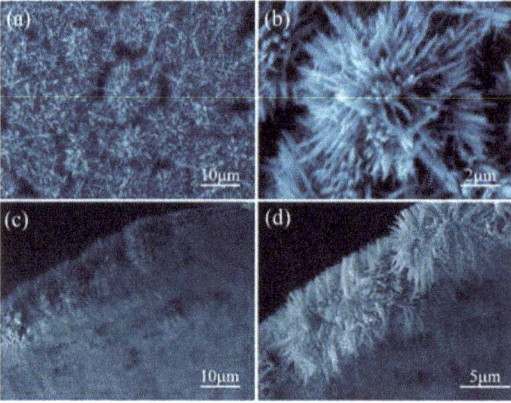

Figure 2. SEM images of TiO$_2$ nanoarrays grown on vertically placed ITO substrate: (**a**) Front view of TiO$_2$ nanoarray; (**b**) A magnified view; (**c**) TiO$_2$ nanoarray cross-section; (**d**) C magnified view.

Figure 3 shows the microtopography of TiO$_2$ nanoarray that places the FTO conducting glass into the reaction vessel vertically at 180 °C for 4 h. Figure 3a,b shows the front view of TiO$_2$ nanoarray grown on conducting glass substrate. As can be seen in Figure 3a,b, there are stacked sea urchin-shaped TiO$_2$ nanoarrays on the surface, with different sizes and dense arrangement. The length and diameter of the nanorods are basically equal in the same urchin-shaped TiO$_2$ nanoarray. The inset in Figure 1a is the cross-section diagram and enlarged view of TiO$_2$ nanoarray generated on the FTO conducting glass. It can be clearly seen that the TiO$_2$ grown on the horizontally placed conducting glass is composed of TiO$_2$ nanorod arrays and urchin-shaped TiO$_2$ nanoarrays that are in contact with the bottom surface. The two kinds of nanoarrays with different morphologies are combined to form the homojunction of the TiO$_2$ nanoarray (recorded as TiO$_2$ –HJ, HJ: Homojunction), and the thickness of the homojunction of the TiO$_2$ nanoarray is between 150 and 200 mm. Figure 3c,d shows the reverse side of the TiO$_2$ grown on the FTO conducting glass and their enlarged views. The reverse side indicates the contact surface between TiO$_2$ nanoarray and conducting glass. As can be seen in Figure 3c,d, the nanorods are arranged in order with the diameter of 600 nm and the vertical extent of the TiO$_2$ nanorod array is obviously improved when the FTO conducting glass is placed vertically.

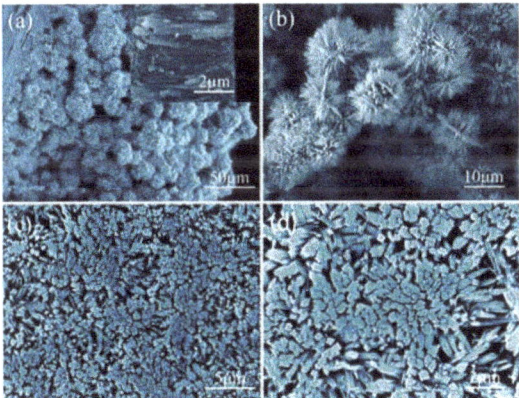

Figure 3. SEM images of TiO$_2$ nanoarrays grown on horizontally placed FTO substrate: (**a**) Front view of TiO$_2$ nanoarray; (**b**) A cross-sectional view and an enlarged view of the TiO$_2$ nanoarray; (**c**) TiO$_2$ nanoarray bottom view; (**d**) C magnified view.

Figure 4 shows the microtopography of TiO$_2$ nanoarray that places ITO conducting glass into the reaction vessel vertically at 180 °C for 4 h. Figure 4a,b shows the front view of the homojunction of the TiO$_2$ nanoarray grown on the ITO conducting glass substrate. There are stacked urchin-shaped TiO$_2$ nanoarrays on the surface, with different sizes and dense arrangement. The morphology of the growth of the ITO conductive glass is similar to that of the nanoarray formed when the FTO conductive glass is the substrate, and the nanorods are arranged radially. The urchin-shaped TiO$_2$ nanoarrays are substantially equal. The inset in Figure 4a is the cross-section diagram and enlarged view of TiO$_2$ generated on ITO conducting glass. It can be clearly seen that the urchin-like nanospheres are supported by TiO$_2$ nanoarrays. The thickness of the homojunction of the TiO$_2$ nanoarray is between 150 and 200 mm. Figure 4c,d shows the reverse side of TiO$_2$ grown on ITO conducting glass. It can also be seen from the back view that the compactness of the TiO$_2$ nanorod array, which is generated by taking the ITO, is lower than that of the TiO$_2$ nanorod array that takes the FTO conducting glass as the substrate.

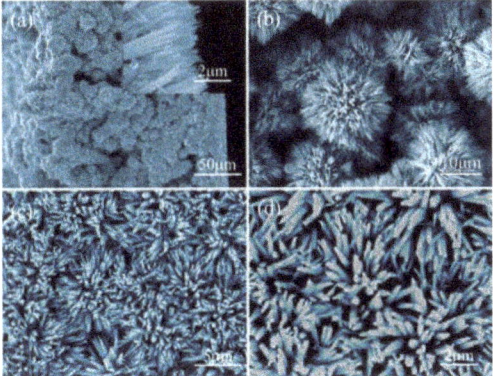

Figure 4. SEM images of TiO$_2$ nanoarrays grown on horizontally placed ITO substrate: (**a**) Front view of TiO$_2$ nanoarray; (**b**) A cross-sectional view and an enlarged view of the TiO$_2$ nanoarray; (**c**) TiO$_2$ nanoarray bottom view; (**d**) C magnified view.

The morphology of nanorods grown on various conducting glasses are different, which may be due to the different crystal structures of the conducting glasses. The ITO glass is composed of 10 wt.% SnO_2 doped In_2O_3. SnO_2 have rutile structure, while In_2O_3 has cubic ferromanganese crystal structure. TiO_2 epitaxial grow on the ITO surfaces to obtain loose nanoarray with rutile structure due to the low content of SnO_2. By comparison, the FTO glass is only composed of rutile SnO_2. TiO_2 epitaxial grow on the external surfaces of the SnO_2 to obtain more orderly and denser rutile nanorod arrays.

Moreover, it can be found that under the same hydrothermal conditions, the homojunction is formed when the conducting glass is placed horizontally in the reaction vessel, while only nanoarray is formed when the conducting glass is placed vertically. Under certain temperature conditions, the TiO_2 generated by hydrolysis nucleates on the conducting glass and continues to grow into nanorods after nucleation. When the nanorods grow to a certain length, the hydrolysis reaction continues to produce a large amount of TiO_2. TiO_2 grows into urchin-shaped-like TiO_2 nanoballs in solution. A large amount of TiO_2 exceed the growth rate of TiO_2 nanorods along the vertical direction on the conducting glass and falls directly on the TiO_2 nanorods that have been already generated, which forms the titanium dioxide nanoarray homojunction. The schematic diagram of the formation mechanism is shown in Figure 5a shows the schematic diagram of the formation of TiO_2 nanoarray when the conducting glass is placed vertically. Figure 5b indicates the schematic diagram of the homojunction of TiO_2 nanoarray when conducting glass is placed horizontally.

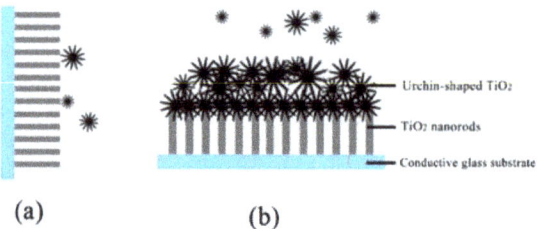

Figure 5. Generation mechanism diagrams of TiO_2 nanoarrays and its homojunction: (**a**) Placing the conductive glass vertically; (**b**) Placing the conductive glass horizontally.

Figure 6 shows the Raman spectrum of TiO_2 nanoarray, which shows well-resolved peaks of TiO_2–FTO–HJ located at 240.35 cm^{-1}, 443.13 cm^{-1}, and 610.42 cm^{-1}, and the well-resolved peaks of TiO_2–ITO–HJ located at 235.29 cm^{-1}, 444.82 cm^{-1} and 610.42 cm^{-1}. From these results, it can also be concluded that the crystal structures of the two homojunctions are rutile [26].

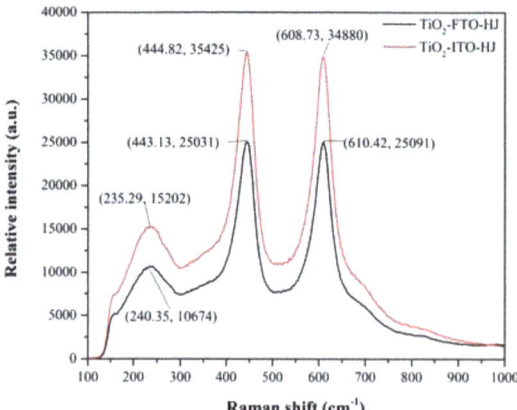

Figure 6. The Raman spectrum of TiO_2 nanoarrays.

3.1.2. The Structure of the TiO$_2$ Nanoarray

X-ray diffraction (XRD) characterization is presented in Figure 7. It can be seen that the TiO$_2$ with four morphologies have crystal faces of (110), (101), (002), (301) and (112), which indicates that the crystal structures of TiO$_2$–FTO, TiO$_2$–ITO, TiO$_2$–FTO–HJ and TiO$_2$–ITO–HJ are tetragonal rutile phase [27]. The (101) diffraction peak of TiO$_2$ grown on FTO conducting glass substrate is the strongest and sharpest, which suggests that the structure order degree of TiO$_2$–FTO was improved by FTO [26,27].

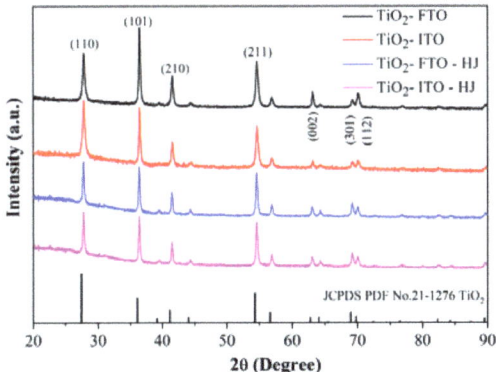

Figure 7. The XRD spectrum of TiO$_2$ nanoarrays.

It can be seen from the FT–IR spectrum of TiO$_2$ nanoarrays in Figure 8 that a weak absorption appears at the wavenumber of 3400 cm^{-1} the TiO$_2$ grown on the FTO and ITO conducting glass substrates. The weak absorption of 3400 and 1600 cm^{-1} is the characteristic absorption of hydroxy (–OH), which may be due to the incomplete hydrolysis of titanium tetrachloride (TiCl$_4$) to generate TiO$_2$ and residual –OH. The characteristic bands at 550 cm^{-1} are attributed to vibrations of Ti–O in TiO$_2$ nanocrystal.

Figure 9a is the ultraviolet-visible absorption spectrograms of four nanoarrays and their band gap diagrams. The homojunction of the TiO$_2$ nanoarrays generated by the horizontally placed FTO and ITO conducting glasses are basically the same; we can see that TiO$_2$–FTO, TiO$_2$–ITO and ITO–HJ absorption wavelength in the ultraviolet region from 400 nm, while TiO$_2$–FTO–HJ has an absorption band edge at the about 750 nm. It can be seen from Figure 9b that the band gap of TiO$_2$–FTO is 3.02 eV, and the band gap of TiO$_2$–ITO, FTO–HJ, ITO–HJ is 3.04 eV, respectively, suggesting that these TiO$_2$ nanostructures may be promising for sensing [28].

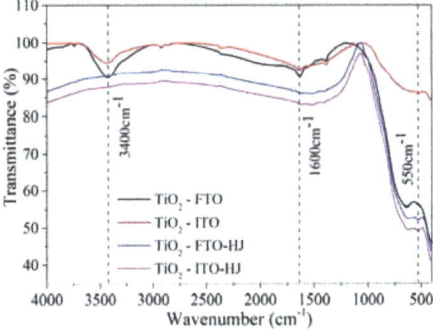

Figure 8. The FTIR spectrum of TiO$_2$ nanoarrays.

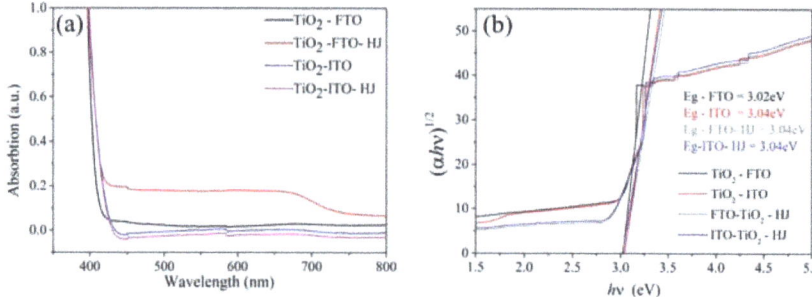

Figure 9. UV-Vis absorption spectrum and band gap calculation diagram of TiO$_2$ nanoarrays: (**a**) Ultraviolet-visible absorption spectrum; (**b**) Band gap calculation chart.

The gas sensing performance diagram of TiO$_2$–FTO–HJ, TiO$_2$–ITO–HJ, TiO$_2$–FTO and TiO$_2$–ITO to TATP vapors was evaluated under simulated natural light [29]. One can see from Figure 10a that the response of TiO$_2$–FTO–HJ to TATP vapors is nearly three times as strong as the response of TiO$_2$–ITO–HJ to TATP. TiO$_2$–FTO and TiO$_2$-ITO also respond to TATP vapors, but the response is weak compared with that of homojunction structure. As shown in Figure 10b, the responses of TiO$_2$–FTO–HJ, TiO$_2$–ITO–HJ, TiO$_2$–FTO and TiO$_2$–ITO to TATP vapors were about 67%, 23%, 11% and 7%, respectively. Moreover, TiO$_2$–FTO–HJ showed a quick response time to TATP of 7.2 s and a recovery time of 4.9 s (Figure 10b). This can be attributed to the homojunctions formation between the urchin-shaped TiO$_2$ and TiO$_2$ nanoarrays, which may be beneficial in promoting the efficient separation of photogenerated charges.

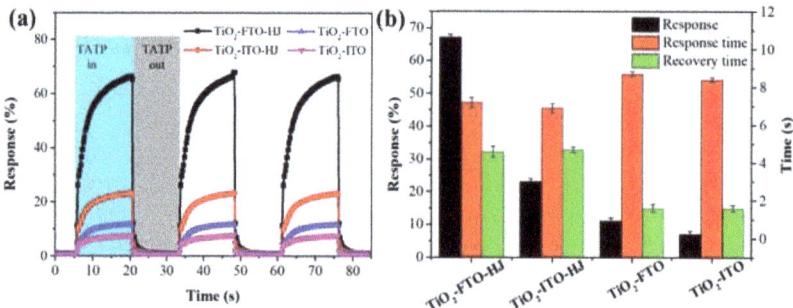

Figure 10. Gas sensing performance diagram of explosives of TiO$_2$ nanoarrays (**a**) and the corresponding statistical chart of response, response time and recovery time (**b**).

As a demonstration, E-field intensity distributions of TiO$_2$ homojunction and TiO$_2$ nanoarrays were simulated by finite element method. Figure 11a shows the electric field distributions of TiO$_2$ homojunction. The hot spot area can be found between the interface of urchin-shaped TiO$_2$ nanosphere and TiO$_2$ nanoarrays. Compared with the simulation results of the urchin-free nanosphere shown in Figure 11b, the strength of the electromagnetic fields of the TiO$_2$ homojunction at the hot spot area increase at least six orders of magnitude with respect to the TiO$_2$ nanoarrays. This demonstrates that the interface of the urchin-shaped TiO$_2$ nanosphere and TiO$_2$ nanoarrays can enhance the localized electromagnetic fields intensity to a certain extent under the simulated natural light. Thus, the photoelectrons accumulate at the interface of the homojunction to provide photocatalysis active site to reduce TATP to acetone and hydrogen peroxide (Figure 11c).

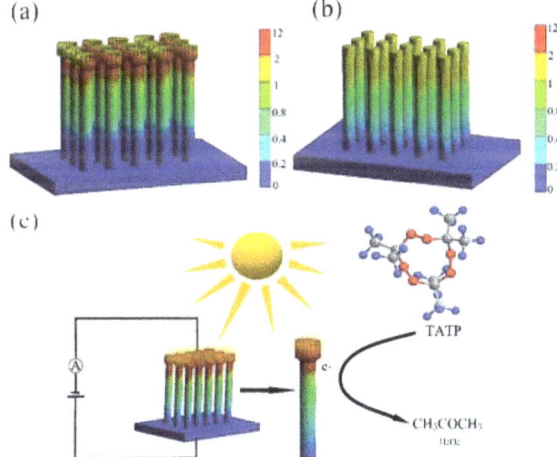

Figure 11. E-field intensity distributions of TiO$_2$ nanoarray on FTO glass: (**a**) Front view of a TiO$_2$ nanoarray applying a uniform E-field; (**b**) Front view of a dandelion-free TiO$_2$ nanosphere array applying a uniform E-field and (**c**) a possible sensing mechanism of TiO$_2$–FTO–HJ to TATP.

Ammonium nitrate (AN), urea and sulfur (S) are also employed as improvised explosives or explosive raw materials. The sensing properties of the TiO$_2$–FTO–HJ toward different saturated AN, urea and S vapors are depicted in Figure 12. The response of TiO$_2$–FTO–HJ to AN, urea and S vapor is about 22%, 26% and 9%, respectively, which is lower than the response to TATP. The response time of TiO$_2$–FTO–HJ to AN, urea and S vapors is about 11 s, 24 s and 12 s, while the recovery time is 12 s, 4 s and 3 s, respectively. Among these improvised explosives, AN, urea and S are relatively stable. However, the electron transfer process is thermodynamically and kinetically favorable for TATP reductive decomposition. Thus, TiO$_2$–FTO–HJ shows selectivity to TATP, and this result reveals that TiO$_2$–FTO–HJ is a potential candidate for TATP sensors.

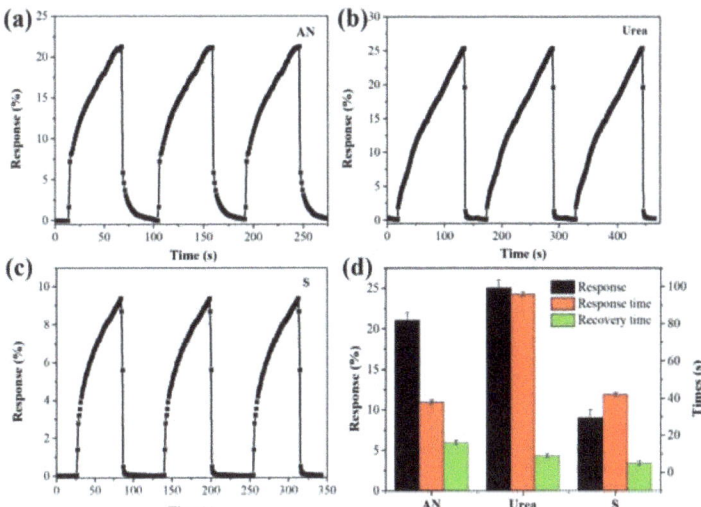

Figure 12. Time-dependent response of the TiO$_2$–FTO–HJ in room-temperature to saturated AN (**a**), urea (**b**) and S (**c**) vapor and corresponding response time and recovery time (**d**).

4. Conclusions

The TiO$_2$ nanoarrays and sea urchin-like TiO$_2$-covered TiO$_2$ nanoarray epitaxial were grown on FTO and ITO conductive glass using a simple one step hydrothermal method. For the TiO$_2$ nanoarrays grown on the FTO and ITO conducting glasses, when they were placed horizontally, the part that was in contact with the conducting glass was more orderly and uniform than when vertically placed. In addition, for the nanoarrays that used FTO conducting glass as the substrate, the nanoarrays were more orderly and uniform than those using ITO conducting glass as the substrate. The crystal structure of the homojunction of TiO$_2$ nanoarrays grown on conducting glass placed horizontally was rutile. Among the four kinds of TiO$_2$ nanostructure, TiO$_2$–FTO–HJ showed excellent response to TATP vapors, reaching 67%, compared with the performances of TiO$_2$–ITO–HJ (23%), TiO$_2$–FTO (11%), and TiO$_2$–ITO (7%) with respect to TATP. The response speed and recovery speed of TiO$_2$–FTO–HJ to TATP were relatively fast, with a response time of 7.2 s and a recovery time of 4.6 s. The TiO$_2$–FTO–HJ also showed high selectivity to TATP among ammonium nitrate, urea and sulfur, which are common homemade explosives raw materials.

Supplementary Materials: The following are available online at http://www.mdpi.com/2079-6412/10/4/409/s1, Table S1: The available techniques for detection of TATP.

Author Contributions: Conceptualization, B.L. and L.G.; methodology, Y.T. and Y.Z. (Yuxiang Zhang); validation, Y.Z. (Youxiong Zheng) and J.Y.; formal analysis, Y.Z. (Yuxiang Zhang); investigation, Y.T., Y.Z. (Yuxiang Zhang) and G.X.; resources, B.L.; data curation, Y.T. and Y.Z. (Youxiong Zheng); writing—original draft preparation, Y.T. and Y.Z. (Yuxiang Zhang); writing—review and editing, B.L. and L.G.; visualization, Y.Z. (Youxiong Zheng); supervision, B.L. and L.G.; project administration, B.L.; funding acquisition, B.L. and L.G. All authors have read and agreed to the published version of the manuscript.

Funding: This research was funded by the National Natural Science Foundation of China 21804078, Natural Science Foundation of Qinghai Province 2020-ZJ-764 and Thousand Talents Program of Qinghai Province.

Conflicts of Interest: The authors declare no conflicts of interest.

References

1. Hernández-Adame, P.L.; Medina-Castro, D.; Rodriguez-Ibarra, J.L.; Salas-Luevano, M.A.; Vega-Carrillo, H.R. Design of an explosive detection system using Monte Carlo method. *Appl. Radiat. Isot.* **2016**, *117*, 27–31. [CrossRef] [PubMed]
2. Sun, X.; Wang, Y.; Lei, Y. Fluorescence Substrated Explosive Detection from Mechanisms to Sensory Materials. *Chem. Soc. Rev.* **2015**, *44*, 8019–8061. [CrossRef] [PubMed]
3. Lichtenstein, E.; Havivi, R. Shacham, Supersensitive Fingerprinting of Explosives by Chemically Modified Nano sensors Arrays. *Nat. Commun.* **2014**, *5*, 41–95. [CrossRef] [PubMed]
4. Pinnaduwage, L.; Gehl, A.; Hedden, D. Explosives a Microsensor for Trinitrotoluene Vapour. *Nature* **2003**, *425*, 474. [CrossRef] [PubMed]
5. Wang, F.; Gu, H.; Swager, T. Carbon Nanotube/Polythiophene Chemiresistive Sensors for Chemical Warfare Agents. *J. Am. Chem. Soc.* **2008**, *130*, 5392–5393. [CrossRef] [PubMed]
6. Mao, S.; Zhou, H.; Wu, S. High performance hydrogen sensor substrated on Pd/TiO$_2$ composite film. *Int. J. Hydrogen Energy* **2018**, *43*, 22727–22732. [CrossRef]
7. Ho, W.-J.; Hsiao, K.-Y.; Hu, C.-H. Characterized plasmonic effects of various metallic nanoparticles on silicon solar cells using the same anodic aluminum oxide mask for film deposition. *Thin Solid Films* **2017**, *631*, 64–71. [CrossRef]
8. Chen, C.C.; Chen, S.H.; Shyu, S.W. Use of Nanostructures in Fabrication of Large Scale Electrochemical Film. *Phys. Procedia* **2012**, *25*, 44–49. [CrossRef]
9. Lee, J.; Kim, D.H.; Hong, S.-H. A hydrogen gas sensor employing vertically aligned TiO$_2$ nanotube arrays prepared by template-assisted method. *Sens. Actuators B Chem.* **2011**, *160*, 1494–1498. [CrossRef]
10. López-Ayala, S.; Rincón, M.E.; Pfeiffer, H. Influence of copper on the microstructure of sol–gel titanium oxide nanotubes array. *J. Mater. Sci.* **2009**, *44*, 4162–4168.
11. Huang, J.-W.; Lu, K.C.-C.; Huang, Y.-S. Novel fabrication process using nanoporous anodic aluminum oxidation and MEMS technologies for gas detection. *Procedia Chem.* **2009**, *1*, 56–59. [CrossRef]

12. Ge, Y.; Zhong, W.; Li, Y. Highly sensitive and rapid chemiresistive sensor towards trace nitro-explosive vapors based on oxygen vacancy-rich and defective crystallized In-doped ZnO. *Sens. Actuators B Chem.* **2017**, *244*, 983–991. [CrossRef]
13. Zeng, X.S.; Zeng, X.S.; Xu, H.; Xu, Y. A series of porous interpenetrating metal-organic frameworks based on fluorescent ligand for nitroaromatic explosives detection. *Inorg. Chem. Front.* **2018**, *10*, 10–39. [CrossRef]
14. O'Mahony, A.M.; Wang, J. Nanomaterial-based electrochemical detection of explosives: A review of recent developments. *Anal. Methods* **2013**, *5*, 4296. [CrossRef]
15. Kielmann, M.; Prior, C.; Senge, M.O. Porphyrins in troubled times: A spotlight on porphyrins and their metal complexes for explosives testing and CBRN defense. *New J. Chem.* **2018**, *42*, 7529–7550. [CrossRef]
16. Meng, X.C. Detection Technology of Explosives and Drugs. *Nucl. Electron. Detect. Technol.* **2013**, *4*, 371–379.
17. Liang, J.H.; Liu, P.P.; Chen, Z.; Sun, G.X.; Li, H. Rapid evaluation of arsenic contamination in paddy soils using field portable X-ray fluorescence spectrometry. *J. Environ. Sci.* **2018**, *64*, 345–351. [CrossRef]
18. Patil, V.L.; Vanalakar, S.A.; Patil, P.S. Fabrication of nanostructured ZnO thin films substrated NO_2 gas sensor via SILAR technique. *Sens. Actuators B Chem.* **2017**, *239*, 1185–1193. [CrossRef]
19. Drobek, M.; Kim, J.H.; Bechelany, M. MOF-Substrated Membrane Encapsulated ZnO Nanowires for Enhanced Gas Sensor Selectivity. *ACS Appl. Mater. Interfaces* **2016**, *8*, 8323–8328. [CrossRef]
20. Wang, J.; Wang, Z.; Li, H.; Cui, Y.; Du, Y. Visible light-driven nitrogen doped TiO_2 nanoarray films: Preparation and photocatalytic activity. *Alloys Compd.* **2010**, *494*, 372–377. [CrossRef]
21. Jie, C.; Lin, C.; Dahai, Z. Surface Characteristic Effect of Ag/TiO_2 Nanoarray Composite Structure on Supercapacitor Electrode Properties. *Scanning 2018*, *2018*, 1–10.
22. Jain, U.; Singh, A.; Kuchhal, K. Glycated hemoglobin biosensing integration formed on Au nanoparticle-dotted tubular TiO_2 nanoarray. *Anal. Chim. Acta* **2016**, *945*, 67–74. [CrossRef] [PubMed]
23. Hu, W.; Dong, F.; Zhang, J. Differently ordered TiO_2 nanoarrays regulated by solvent polarity, and their photocatalytic performances. *Appl. Surf. Sci.* **2018**, *442*, 298–307. [CrossRef]
24. Cheng, K.; Cao, D.; Fan, Y. Electrodeposition of Pd nanoparticles on C@TiO_2 nanoarrays: 3D electrode for the direct oxidation of $NaBH_4$. *J. Mater. Chem.* **2011**, *22*, 850–855. [CrossRef]
25. Zhang, R.; Shao, M.; Xu, S. Photo-assisted synthesis of zinc-iron layered double hydroxides/TiO_2, nanoarrays toward highly-efficient photoelectrochemical water splitting. *Nano Energy* **2017**, *33*, 21–28. [CrossRef]
26. Shinde, D.V.; Mane, R.S.; Oh, I.H. SnO_2 nanowall-arrays coated with rutile-TiO_2 nanoneedles for high performance dye-sensitized solar cells. *Dalton Trans.* **2012**, *41*, 10161. [CrossRef]
27. Wang, H.; Bai, Y.; Wu, Q. Rutile TiO_2 nano-branched arrays on FTO for dye-sensitized solar cells. *Phys. Chem. Chem. Phys.* **2011**, *13*, 7008. [CrossRef]
28. Yang, Y.; Yin, L.C.; Gong, Y. An Unusual Strong Visible-Light Absorption Band in Red Anatase TiO_2 Photocatalyst Induced by Atomic Hydrogen-Occupied Oxygen Vacancies. *Adv. Mater.* **2018**, *30*, 1704479. [CrossRef]
29. Lü, X.; Hao, P.; Xie, G. A Sensor Array Realized by a Single Flexible TiO_2/POMs Film to Contactless Detection of Triacetone Triperoxide. *Sensors* **2019**, *19*, 915. [CrossRef]

© 2020 by the authors. Licensee MDPI, Basel, Switzerland. This article is an open access article distributed under the terms and conditions of the Creative Commons Attribution (CC BY) license (http://creativecommons.org/licenses/by/4.0/).

MDPI
St. Alban-Anlage 66
4052 Basel
Switzerland
Tel. +41 61 683 77 34
Fax +41 61 302 89 18
www.mdpi.com

Coatings Editorial Office
E-mail: coatings@mdpi.com
www.mdpi.com/journal/coatings

www.ingramcontent.com/pod-product-compliance
Lightning Source LLC
LaVergne TN
LVHW070604100526
838202LV00012B/562